Worlds Beyond

((Uncanny Science Fiction Stories) (Illustrated Edition))

By, Dr. Dennis L. Siluk

Andean Scholar, and Poet Laureate in Peru

Front Cover and Illustrations by the Author

Worlds Beyond

((Uncanny Science Fiction Stories) (Illustrated Edition))
November, 2016 Copyright © By Dr. Dennis L. Siluk

ISBN-13: 978-1539793403
ISBN-10: 1539793400

This is Dr. D.L. Siluk's 65ᵗʰ Internationally Published Book
(Not including his 48-published chapbooks)

Author Website:
http://dennissiluk.tripod.com

Dedicated to Little Jack Hageman

Who died 3-5-2016 (15-months old)

—and to My Wife, Rosa

This is a work of Science Fiction

"The Paris Book Festival" 2016, Honorable Mention, for a Spiritual Book, completion went to: "The Galilean" by Dr. Dennis L. Siluk.

Recognition given by the Congress of the Republic of Peru, from "DestAcados" Magazine, to Dr. Dennis L. Siluk, for "Promoter of the Culture of the Mantaro Valley of Peru" May 2016.

"You are a good student of human nature and you express it well."
—Janet French (2-25-2016)

"I have been re-reading some of your poems in 'The Galilean' and really enjoying them. I love the one 'The Eternal Present' about God living outside of time. I also enjoy the ones about the saints, especially St. Catherine..."
Gail Weber, Editor / Owner, "Exploring Tosca"
A Minnesota Art & Cultural Magazine, May, 2015

Appreciation by Pope Francis, concerning "The Galilean" through Letter sent by the Apostolic Nunciature. Lima, Peru, 2013.

"The book, 'The Galilean' (Volume One) from the first poem to the last is a story of Salvation..."
—Father Marcelo Loaiza (Parish Priest, St. Daniel Comboni of Christ Redeemer) 2013

"With great regard and admiration...you being a decorous person—: 'Poems for the Soul' (The Galilean) tells us, the reader once more of your endless layers and poetic productions, which exalt the intelligence and human culture."
Dr. Adolfo Vargas
Mayor, San Juan de Miraflores, Lima, Peru. 8-2013

Index

WORLDS BEYOND

THE HIDDEN SCROLLS OF MARS

MARS

((EARTH TIME, 100,000 B.C.)(PART I OF II))

"I hope, nobody finds these writings! If someone does, it means somebody else between Earth and Mars or some other planet has been here, and God help them if they have no way off this planet. That is to say, they'll most likely have to face these awful solar winds and storms which you will find out are quick and intense. They have stripped away most of Mars' atmosphere, edged rock formations, rivers. I am the last person left here, most of us left, escaped to Earth a decade ago. Earth is quite primitive to our race yet it is rich with life. We look more like the Cro-Magnon—according to my comrades that have landed on Earth, and only time will tell—I venture to say—if they will produce a more enhanced stage of mankind.

"From my radio reports, one of our spacecraft had an emergency landing on an asteroid caused by much cosmic dust, perhaps from a hot aftermath of a supernova. What their fate will be, time will tell.

"I am Commander Mruts, I stayed behind, voluntarily stayed behind that is, for the simple reason someone had to man the last of Mars' stations, and monitor what needs to be monitored. Although I get updates from Earth and the Asteroid, to which one of our parties landed on the second part, or larger section of land on the awkward asteroid.

"They have now been gone five years. I am writing these scrolls out for posterity's sake, —in case anyone may fall upon them in the future; I am writing 1800-feet below the surface of Mars, in what I consider a dreadful remoteness; I do understand our astronauts, and those civilians on those spaceships were hurled through space like a bouncing ball, it turned out to be a nasty and horrific trip, with a comet nearly shifting the vessels' courses as it orbited by.

"The Asteroid, will not allow the Martians to live much longer I am told, their supplies are down, as is their fuel, and there appears to be no escape, or launch ability; on the other hand, here I am to die in this planetary pit, without demur.

2

"We call Earth the blue planet, and Mars the red planet. At one time Mars was shrouded in thick gases which supported our presence, and we had much water on the surface, many rivers. Like Earth, too, at one time the presence of air was plentiful, I mean its content was antiquated, our CO_2 level & O_2 level within the atmosphere was like Earth's is today ((the oxygen within the air consisting of 21% of the atmosphere by volume) (a diatomic gas)), today on Mars that number is less than 5% of Earth's, and our water supply on the surface is completely frozen solid. And the atmosphere is so thin, being removed from high, to higher altitudes yearly; which through interactions with the sun, forced us to build underground domiciles, which supply most of our daily survival needs.

"On the other hand, as I have been informed: Earth has a global magnetic field that continues to operate and protect the life within its climatic system, but I suppose it could in time be altered in a very similar manner likened to Mars. Should their moon be knocked out of its orbit, or Earth off its spin, that is, axis, or an asteroid half the size of Pluto tumble outside its belt, sideswipe it, and move it a foot here or there, consequently, their moon protects them quite well from the sun, and its orbit; yet at present, makes it favorable in that it lights up the night, gives shade to the earth daily, how fortunate, God has been gracious to the earthlings.

"On another subject, my comrades say they must mingle with the Neanderthal, which we call their primitive race, and what we will get in the future is any man's guess. Perhaps we'll all die out, and God will take a few of our DNA, and their DNA, and implant them into a new kind of man! Who's to say?

3

"Again I stress, I hear oxygen and carbon dioxide are plentiful on Earth, this is important as is water for the planet's life overall, which keeps this balance between mammals and vegetation and Earth's overall flora.

"I must say, alas, our magnetic field is crushed, so there is no coming back to Mars other than for old time's sake, and so here goes 4.5 billion years of history, to a nearly unusable, unstable, unpredictable planet, other than holding down a station for whatever purposes. I mean we are 33.9-million miles from Earth, and 141.6-million miles from the Sun, kind of lost in the Gobi Desert of space. So, one of our team members have named one of Earth's deserts.

"I hate going out onto the surface, hard to breathe, and the temperature is minus 60 degrees Celsius at present. But reader, do not dismay, I volunteered for this, someone had to."

In the year 2050 A.D., these ancient scrolls of Mars were found during an exploration of the planet by none other than an old space dog and his astronaut master, Staff Sergeant Hector Hecker, of the U.S.A Independent Space Continuum, (privately owned by Donald Rump, a billionaire), whom was from St. Paul, Minnesota; the dog, said to have been weather-beaten, went by the name Max III; Max's great grandfather fought in the Afghanistan War, and thus, Max III is now in the annals of Interplanetary Space Exploration Museum. As for the Staff Sergeant he didn't do the finding, and well, —dogs have their rights too.

#4919/11-21-2015 (Reedited and revised slightly, 12-28-2015) / 11/2015 See Part II, "Last Voyage from Mars" Reedited 8-30-2016.

Last Voyage from Mars

((Commander Mruts) (Part II of II))

I am caught within a trans-dimensional vortex, carried by a gravitational wave ((GW) (inside the wave's pocket)), inside a semicircular vessel; furthermore, caught in the pull of the wave at the speed of light. I will never age at this speed, and now the wave just passed Earth, I fear I shall be on my way to the end of the universe, what end is anyone's guess. Who's to say, perchance this was the very wave that created the Big Bang after matter and antimatter collided, some 14.5 billion years ago. Or perhaps it is some wave that was created a billion years ago from some supernova, or two black holes with their horrid gravitational pull, plunging into one another and creating a giant blast. But here I am all the same. It would seem I was asleep a long while, perhaps this vessel was pulled off its course on my way to Earth from Mars, like snatching a hitchhiker off a highway at full speed; thus, my vessel adjoined with the GW, in what one might refer to as transcendental, or awe-inspiring space. Now lost in its supernal realm, for how long, no one knows!

Nothing was clear, everything obscure in my head, evidently I was in my space station on Mars, had secured the spacecraft, took somehow a head injury along the way, the concussion has not allowed me to put this puzzle all together, and this vessel got caught in the gravitational wave's way. I must have purged through an interspatial vacuum, with a gravitational wave plowing through it at the same time, and taken me off my course to Earth. I do remember working on this super-aqueous apparatus for the vessel that would produce water indefinitely, and if need be, for an endless supply of water on Mars. I was deep in my granite cabin of sorts, or lab, on Mars, the vessel was ready to go, and I must have hit the GW, and I now just awoke, or a few hours ago. So, I can't say how long I've been caught in this situation, only that I am in this interval of time, what I suspect happened, most likely did happen. I do remember it all was experimental, in the underground lab on Mars. I was hoping to be thrown into Earth's outer rim of space, but the wave this entity that carried me, pulled me, my vessel away from its goal. How unfortunate.

Time Passes

More and more I am heading to who knows where. I have passed Jupiter, Pluto, long ago, way beyond them now. I can count moment by moment, but not real-time; once I get out of this enfolded wave, I'll feel better, I sense I am in the belly of the whale, figuratively speaking, yet in reality I am in a swollen part of the waves quarry. And the wanting of my brain to return ungraspable images that I can understand before I made this adventure, but much is still a blur. I sense months and seasons are passing me by as I travel, and I never seem to get a day older. Time does not pass at the speed of light, it is obliged to follow the laws of physics, but perhaps God has some unknown laws I've yet to learn, laws no one knows. I am now awaiting a miracle, those laws of physics only God knows. How old will I be when I find one? My friends who left Mars for Earth, eons ago, I'm sure all are dead by now, perhaps they have become part of the human race, what was then an unhospitable environment, and that is to say, in perhaps passing millenniums have taken place.

Time Passes

Earth looked to me like an opal last time I saw it, it was then, 100,000 B.C., on Earth, my ship reads now it is 2020 A.D., I presume the wave somehow turned, it must had been twenty-lifetimes I've been in this ship, a tomb of sorts, it is like a time machine that has no time to register, leaping through space like a jewel, across the universe, like a colossal tidal wave (tsunami) then somehow the GW drooped down and I beheld the face of Earth, and now my ship is falling, falling. I've noticed, how swiftly the curvature of the wave swelled and that swelling has pushed my vessel out of its pocket—thank goodness, and now I am falling, falling: descending to the earth, into its embowered sea, called 'Red'.

Now I'm on the surface of the earth, I must start a new life; I wonder how old I am in earth years? Who can I tell, where do I go, and what next should I expect? I think I'm in an Ice Age?

#5071/2-13-2016/reedited 2-14-2016, redrafted 2-20-2016 / See part one "The Hidden Scrolls of Mars" Reedited, 5-11-2016 & 8/2016

Note on the Planets: Venus and Mercury, the two closest planets to the sun, and the hottest; Mercury being a little larger than our moon, no wonder it is burnt black, has no moons to shade it, as does Venus have no moons. Unlike Earth, whose moon is nearly as large as Mercury! Mars has two small moons, thus it has a thin atmosphere, being the fourth planet from the sun, and quite small compared to Earth, Earth being the third from the sun.

Note: Light travels at a constant, finite speed of 186,000 mi/sec. A traveler, moving at the speed of light, would circumnavigate the equator approximately 7.5 times in one second.

THE DARK PLANET

(With Dr. G. B. McGee) Part I

"When the Dark Planet that we have recently called an Exoplanet (or extrasolar planet, a planet orbiting a star other than our sun), when it moves it is moved by the neighboring galaxy's gravity and there within its explosions," explained Dr. G.B. McGee, to the committee, at the Louisiana Space Station, adding "in addition to other forces I will explain later. But what I want to say first and foremost is we are reading events that have taken place in space and time by detecting gravitational waves, called warping of time, space produces cataclysm, and now we can read events that have taken place millions of years ago—waves are like natures, or environments, inside their DNA—((figuratively speaking) (throughout time, can track their way historically, to and fro)). And what appears to have recently taken place, perhaps has taken place three-million years to get to us, so it is all dead data. Again gentlemen and women, I shall get back to this sooner than later. What I've noticed recently is hot material by that Dark Planet, we call an Exoplanet discovered by the NASA's Kepler telescope, with one star. This material I am talking about is 'blow shocks' and as they travel, this material piles up in front of them, warms up and glows, normally this is what lights up runaway stars, because it is hard to see them from our telescope, but these blow waves are infra-red, and can be detected. I've recently noticed them by the Dark Planet. For some odd reason, which again I will talk more about later, they have entered the pathway of the Exoplanet. They look like bits and pieces of a nebulae that track stars, and have been redirected into this area I thought was a forbidden zone not too long ago, a year or so ago; this attraction, is by the Dark Planet's sun, apparently by plowing of a supersonic wave of gas.

"Also what I will be telling you in a moment is another intriguing predicament—and perhaps that's too harsh of a word, but solar winds create bow waves, which in essence brighten up their fate—and in time these runaway stars explode, as a supernovae and go down to rest in their gopher hole, we call a black hole. These waves come and go and light up the Dark Planet, then are blown off-course and out of its region, flung into other gravitational forces in the jam-packed stellar neighborhood beyond Exoplanet's black matter, or antimatter.

"But the reason I called this emergency meeting is on another matter: I've just discovered a new black hole! Well not quite, remember this took place perhaps 1000-light years away, three million years ago. I fear I'm going to see in the next projection of our gravitational waves, the disappearance of this dead zone, and this Dark Planet..."

"Why is that," commented Professor Hightower, "I hope you've got a good answer, because I was in the process of flying to Dieburg, Germany, for a reunion with old friends a minute before your phone call...." Professor Hightower, being the Station's Superintendent.

"The black hole I discovered is starting to burp up all its hot gases, and is creating a whirlpool next door to the Dark Planet's solar system. And it is forming a new Galaxy," continued Professor McGee, "this new whirlpool galaxy will take all my attention. I fear the Dark Planet is at present about to disappear, to be swallowed up, and thrown upside-down and into whatever kind of nebulous debris black holes produce, then thrown out, and throughout the new Galaxy in a new kind of creation, gone like a stray dog, that comes back howling like a jackal."

"We'll then see, I do believe" said Dr. Hightower, "the forming of a new galaxy is in the making, or so it sounds!"

Butch McGee, my old Army buddy, looked torn with that comment. For myself, I went to see Father Marcelo after the meeting, but he was at a spiritual retreat; so, I talked to Father Washington, his colleague, who was as much a philosopher as he was a theologian. And he said in a modest term, "God is simply showing you how black holes can have a reversed nature; are they not known to be destructive! Now you see how they can be creative. God is showing you by opening up a new door, or both sides of the coin."

Then he added, "I have a baptism to do, I'll stop over for coffee later if you don't mind, as long as you've got a lot of sugar!"

"Of course," I agreed, "please do."

#4981/1-6 & 7-2016

Space Mining on Asteroid

13,499! (2025 A.D.)

It has been said by nearly all of the folks I know, friends and associates in Minnesota, that my pursuing this space mining adventure, the first of its kind, that I must have been a little touched in the head: that surely some grave disaster is bound to happen, and it is happing as I write this narrative, or journal, to which may very well be my last. But I have proved it is possible, the once thought super-scientific madness of mining of an asteroid in space, at one time was strictly Science Fiction, but today it is now non-fiction, doable.

But here I am, doing just that. I have invested in some companies, two decades ago, did the legal framework in the United States, and went about my business. I have a team of several engineers with me, and two scientists, and there are these dreadful creatures at night, unknown entities that circle this asteroid. I was sane at the outset of this adventure, perhaps not so much anymore, but sane enough to write down in a sober enough way this lucid account of our last several months work, and tonight's predicament.

This is no Jules Verne chronicle, but basic new technologies. We landed several months ago on this asteroid, and we are mining it. Matter of fact, we have some European entrepreneurs backing this project up, along with American Companies, and we've found valuable resources in this so called space body.

We're not sure if we are in violation of the UN's Outer Space Treaty, but it is of less concern at this point, and to be taken up at a later date. And as far as we are concerned, the treaty excludes the appropriation of materials found here. We are like fishermen in international space as the term may be inferred on Earth to: or better yet, in international waters; indeed we are searching the oceans and as well as space, for the same material, and in both cases, the raw materials are deep rooted.

We are on one of the largest of the 13,500 asteroids known to the specialists on Earth, concerning Asteroids; I have named this one: Asteroid 13,499, which is one of the so-called near-Earth asteroids. On this giant asteroid, we've found water, for our in-space activity. And we have a large surface abundance. And we found a source of food, a beast-sea serpent of sorts, called a 'sock' (its official name being: Xenoturella), it has no eyes, teeth, brain, it sucks in its food through a hole, for a mouth, but has no stomach, it is about eight inches long, tastes like

14

raw eel. One of the scientists said it is a very ancient creature. I don't know its chemistry, but when you're hungry who cares. So, on one hand we found the Holy Grail of space, on the other, well, it's a different story.

So, what have we found thus far for what we came for? Our mine is high on platinum-group metals, to include: Iridium, palladium.

However, I wish to bring to the reader's attention, if time permits, and we are still alive, it will be because we've found a way to kill these strange visitations from these phenomenal and deadly creatures: but I am not sure if that is my aim, and at this very minute, I think it's a tinge too late.

There are insects that fly about without wings, at night like bats, like bull-mosquitos, if they bite you it is like hemlock being injected into you. Of the several, three engineers are dead thus far, and as far as us scientists go, all three of us are still alive. These insects are quite the intimidating kind. Their center or hub is behind one of the hummocks on the Asteroid. And we've discovered when they land, or somewhat perch, they become mighty jumpers, the larger ones in particular. I just discovered that the other night, as I stood there for a long while gaping at one of them. It beheld for me a marvel beyond understanding, they can jump several feet, and breathe out an unearthly odor. And they murmur with their breathing. My main emotion is at present, a half-mystic wonder, a deadly curiosity, and the mining is taking second place. My mind is spun dizzily, wanting to capture a few and bring them back to Earth to study, I should say that that now has passed into oblivion. I am for the most part, half-fearful and half-exultant. And to kill them, I have no plausible clue, other than, stomp them out with some sort of insecticide, that we'd have to create with the materials we have at hand, within the compound. And I don't want to be too hasty. We could disinfect them perhaps, by approaching the top of the mound just above them, and letting some substance descend on top of their nest, hence, sterilizing them of their toxic venom, but I suggested we do naught. Other than that, I have done some sketches of the creatures.

The Asteroid Bull-mosquitos

They can be sought out and surveyed in the mornings cloudless light, also, they are like friable soil. And recently they have hidden in sheltered spots, knowing I have been seeking them out to study their behavior; they particular lurk behind giant granite fragments of the asteroid's rough surface.

For various reasons I have dismissed telling our agency on Earth of these bizarre asteroid creepy-crawlers that fly, jump and bite. This enigmatic breed of nocturnal manifestations of killers, about the size of my fist, the largest. To my utter dismay, I shall keep it a secret a while longer. They have become familiar with me, and think I am a less concernment to them than my team members, and should they all be dead in the morning—my team members—which is not so absurd as one might think, I shall live among them if possible by some incalculable means.

I can survive among them I think. I sense the test will be soon, and not unimaginable, they have already this evening started to circle our encampment. I think for once in my lifetime, my instincts are totally right on, they will attack. Thus, as I write this out, I could warn everyone, but if I do, they'll kill me along with them, that's another not so unimaginable instinct. And so, my warning them is unapproachable to my thinking.

Oh yes, there the jumpers are, overhanging the bed nets of the few engineers left and the two scientists, I ponder to think if they can gnaw through those nets, it is curtains for everyone. These are the infinitudes everyone on Earth failed to explore, these ungovernable insects.

#5052/2-5-2016

17

Black Moon Widows

(The Kerberos Narrative/ 2048 A.D.)

Kerberos, orbits Pluto some 36,660-miles from Pluto's alignment with its other moons, to which Pluto has a total of five moons, and in comparison in distance to the Sun: Earth being 92.969.000-million miles from the Sun, Pluto, is 4.67 billion miles away from Earth. But Kerberos is where our story leads us.

It is the most obscure of the five moons I do believe, all being nearby the Kuiper Belt. Kerberos' largest lobe is but eight-km across and its smaller lobe being only five-km across, large asteroids can be much bigger.

Because of its isolation from the other moons, and its small size, asteroids being larger as I've already mentioned, this dwarf moon has a weird inhabitants, — and perhaps for good reason was selected over the other four moons, and not by accident. They harbor the Black Moon Widows (spiders as large and some larger than those old iron frying pans my mother used to use when I was just a lad of ten; you could cook a whole chicken in that frying pan, it weighed six-pounds, no kidding).

The moon is so infested with those shrill and eldritch spiders they constantly are bumping heads—plus there is no gazebos, or shelters for them, they live out in the open, infallible, or foolproof to its weather conditions, so the last report read coming from the spacecraft, Enceladus, named after Saturn's moon for whatever reasons; and from there, via, to me at the Space Center in Louisiana.

THE NARRATIVE

The moon's light indicated to those of the spacecraft Enceladus I (Earth Time, 10,000 B.C.), that it was to a certain degree a living moon, and thus, provided an atmosphere, thin as it was, there was one. Although the moon has darkened over time, a result of chemical changes triggered by sunlight, cosmic rays, and the mass of those Black Moon Widows that cover the moon like black flint, similar to Mercury; thus, those that have survived the initial invasion have multiplied in the ten-thousand years since Enceladus I, when they dropped those organisms off, to its astronomical figure of one billion now, and now being the year 2048 A.D.

Let me say, and in saying this get ahead of myself which will do this account no harm, that the Widows, they will kill their male mates for a song and a dance, and when hungry, are a strain that contaminates wherever they are, whatever they touch, and like a hard-shelled cockroach, hard to kill, and therefore could not be allowed to remain on the far-off planet beyond Earth's solar system by the crew members of the spacecraft Enceladus I.

It might be of interest to the reader, this spacecraft that was being monitored by the Louisiana Space Center Enceladus II, now in the year 2048 A.D., came from what is referred to as the Dark or Black Galaxy; via, a wormhole (or gateway) which can cut a trillion miles into a million one example might be: gravity warps space and time, twists it; in other words, one must find or create a dent in space, a curve, this curvature can be detected, as a result, instead of going around it, which is three times the distance than going through it, you've saved 66.6% distance and time, one must use Pi; time and space can be intertwined likewise. The world is not flat and neither is the universe. And for this story we must take it beyond our imagination and mathematics, and beyond the normal dimensions of space as we know them to be. The planet, the spacecraft had come from, is referred to by its acronym: SSARG, which has two moons.

That said, it was horribly ominous for the Captain of the Enceladus II, and the spacecraft's crew, after 10,000-years down the road to go back to that moon, like opening up a can of worms, or like opening and closing of a funeral, long forgotten; lo, —what did they expect to find? Other than, spider corpses. I mean, we have a 10,000-year separation, a huge gap in time, and one thousand spiders (dropped into what they thought was an endless well, an abyss of no return, that was really a deep pitted crater, likened to a gradated cliff) had turned into one billion in that gap of time. Interplanetary breeding that was not the plan, but rather for the spiders to die— that was the strategy, and they didn't, perhaps one hundred escaped, and colonized the moon, by cannibalization imaginably.

Had you asked the Captain of the spacecraft, Enceladus I, 8000 B.C., that this tiny moon would contain a billion humungous eight legged, two fanged and two feeler, poisonous creatures with loads of cocoons and cobwebs lying about, he would have said, "It's not believable" but we must now stick with Enceladus II, and the year 2048 A.D.

The Enceladus II, had returned on what might be called a routine trip, to investigate this tiny moon, for future possibilities. At this juncture, the spacecraft's captain knew nothing of its inhabitants, blotted out by nearly all light, and only shadows reflecting of the moon's higher or more pronounced geological structures, such as craters and mountains, and gorges. Hence, the Dark Widows covered per near every inch of the moon, making it even darker, as if the terrain was of some dark soil, or rocky material.

With the knowledge of the previous voyage, and the rock hard surface that it once was, —they landed all the same, having no fearful imaginaries. Then, suddenly and per near instantly there came a clambering on the spaceship's outer metallic surface, a horrible droning and hammering, it brought a chill to the crew, the now cringing astronauts, and a deadliness that overwhelmed them, hung over them like a noose cobweb.

Henceforward, the Captain seeing only the dark mass as gravel outside his porthole, weapon in hand, life support suit on, he threw open the door—and the Widows jumped several feet in the air, the Captain pulled the trigger on his shooter, once, twice at a point-blank range, he couldn't miss, but the bullets were pointless with the onrushing gush of Black Moon Widows, for they did not waver, though two of them bled an unknowable white-fungus, the rest unwounded continued moving diabolically throughout the space vessel. Pressing so closely together among the crew there was no room for effectual resistance.

The Captain, no more than a skeleton hanging onto the door hinge, like a thread, he had been the first born of the feast of the moon widows.

Beyond its landing site, the spacecraft was never moved, it was as if to remain in Limbo, another ten-thousand years.

#4926/11-22-2015 (Reedited and revised slightly, 12-27-2015)

McGee's Black Eden

(With Dr. G.B. McGee) Part II

Gateway to the Gods, in Puno, Peru, 17,000 B.C.

Like a dreamer in a dream— Dr. G. B. McGee had come upon a portal (in space and time) actually a doorway, that was a gateway within the new forming Spiral Galaxy. He had been searching for his mysterious Dark Planet that seemingly got thrown about when the previous Galaxy's black hole burped up all its insides, in forming this smaller and more balanced Galaxy. Thus, this undiscovered world was newly discovered again, but things were different this time. Still at the Louisiana Space Station, 2020 A.D., while in the room with the giant telescope, this so called gateway, would open up expressively to permit his entrance, should he have had, a key made out of granite. Then all he needed was step into the doorway, and beyond the mouth of the aurora, into a new dawn. That is to say, allow himself to be swallowed up into its gorge of some monstrous mysterious unknown depths.

The Gateway was a stone structure called "Gateway of the Gods," about forty miles outside of Puno, in the Andes of Peru, built by aliens 17000 B.C., to which he had a key made to fit the impression at the stone door.

Preceding, he figured this ongoing exploration of the Dark Planet had excessively absorbed him night and day, and here it was at his disposal, as if God gave the Universe a mind of its own, and he was allowed to share its mysteries, or on the other hand, was this some game some prankish demon, alien was playing, to whom would open this door, to fall into hell's abyss? After that last thought his body fell into a black panic as he twisted the key to the 'Door of the Gods'.

Was this Dark Planet—seemingly, that no longer was dark, some outpost for an alien race, perchance? So, his inner-soul questioned. The mind can trigger many imaginations when it has no genuine feedback. Therefore, he fumbled some, tottered a bit, and secured his footing as not to fall accidently into the portals mouth, once the door opened. Ere, the portal turned into a rough-arched passage which pushed out a warm aromatic wind, he breathed it all in, it came right from the planet, as if some electrical magnetic forces were in play. Dazzled and bewildered, he had found the Dark Planet, but according to his readings with the gravitational waves, it was 1.5 million years from the previous reading that went back beyond three million years. What should be his next reading should he stay, or if he was to step out of the gateway onto the new planet, would he be the new world's first Adam, or just a lone survivor who took it upon himself to explore other worlds, and perchance get lost along the way?

He no longer saw the original planet's aging dysphonic sun, but a fertile solar system around a giant sun, —in measureless space. The planet although empty of human life, —so he pondered, it looked from where he stood, gorgeous with foliage, although he could see fiery air, here and there. This was his Dark Eden. Weird, yes, and again was this a work of some jealous demon? Hoping he would intrude, thinking he would yield to his craving of discovery. Plus, as he pondered, it came to mind, how would he get back? (Gateways don't always have a way back.)

To retrace that moment in the portal's vortex, was like a living substance. Its glow dropped around him, then unable to resist, his appetence, his craving and desire took over his whole being and he allowed himself to be devoured, swift and warm he fell through the portal of space and time that nearly suffocated him, then falling as If from an airy height, he was on the surface of his Black Eden his once lost and dark planet. (One might even compare it with Lucifer's fall from the cliffs of Heaven, for a nine day drop to Earth, which would make heaven 144-billion miles away). Now alone, he stood stone still, without even an ominous echo from Hell or Heaven, and how far was he from Earth? Trillions of miles? Who's to say!

He trembled 'What to do now?' he said out loud.

He peered fearfully about, it was rich with foliage as he had previous glanced, and the atmosphere was fresh and clean, he could survive and build a shelter, but he had goosebumps, something was wrong, very wrong, and his body knew it before his mind could digest it. What was it? Then he noticed some of the plant life, flowers with big shaped ears like cones, they heard what he said, —and was in some sort of metamorphism translating it within its biochemistry; he knew this as one knows the devil is standing behind you. He made a few more utterances, and the ears moved, and then he knew for sure, they felt he was a trespasser!

On second thought, as days passed into months, whatever portal opened up for Dr. McGee, it now gave him more food for thought: there surely would appear sooner or later the opposite sex, why else would he have been brought here, and not some other place. And he knew his key genes in his immune system was ancient, in that he was one of the 3% of humanity that carried the genetic DNA from the Neanderthals, of 60,000 B.C., which had the ability to fight disease, this could play a key role in the immune system of tomorrow for Dr. McGee's Dark Eden. That is to say, at some point in this world's history, it would be advantages to have a fighting-off source for infections or lethal pathogens from an ancient source, whereupon, on this planet, there was no human life that he knew of yet, but hence, his DNA would help or put up a frontline of defense against bacteria, fungi and/or parasites. The problem now was to wait and see. And hope in the meantime, those large ears of flora could not outthink him, and if possible, make some kind of pact, or treaty or concordat with them, that he was harmless, and a scientist, not a demolisher, or of that sort.

#4983/1-7-2016

THE SEMPITERNAL WORLD

((...OR, 'THE PRIMITIVE AND INESTIMABLE WORLD') (PART TWO TO: "MCGEE'S DARK EDEN") (A SF VIGNETTE)) PART III TO MCGEE SERIES

Dr. McGee, seemingly lost on this new sphere, if not seriously trapped for the moment, which he ordained the Dark Planet, after its Galaxy burped up all its guts, and gases up and out of its black hole to form this new world and tiny galaxy was at this point not a happy camper, not completely, having been on this Sempiternal World for a year.

On the 366th day, his mind stirred when shadows appeared in the far-off distance: 'Are they friendly shadows?' he questioned his mind, and it answered: 'They are, and they are female shadows!' and 'Primitive or civilized shadows,' whatever the case may be, at the moment, they looked like human beings, yet still only misty shadows in the distance, and all the better if his mind was accurate.

Then he heard their voices, female voices; there is a strange warmth to those voices he invoked, and per near a fervor developed, he dropped his fruit, devouring it no more. The closer the voices came, the more aroused he became. He now moved branches out of his way, the sun blotching his vision slightly, shaking his head, putting an elephant type leaf over his head for a hat, he recoiled his vision quickly, he waved madly as to be seen by the shadows, in the far-off distance. They are females for sure he concluded, young and lovely. Without pause, he kept waving his hands in burning contentment. The closer they came the more primitive they looked, on this infinite and incalculable world.

(Interlude: McGee would learn in time, there was a village of women taken from Earth during the 14th century, and these were the remnants of that village, for most all died of the plague but those now who were the approaching shadows, —were the survivors! Brought somehow to this invaluable world through the 'Door of the Gods' his gateway, by whom? Who's to say? Brought to this location by alarm, yet this was a new twilight for him, perhaps the reason he was brought in the first place, if not for his DNA alone, then what for: to examine their primitive intelligence and somehow accomplish in years what would take centuries: to upgrade their heritability, that is to say he knew by evidence people are more or less all the same, intelligence is a giftedness by God, of a people within a civilization, which is a better answer than determining a race is less intellect because of its race, but rather because of its advancement within that race. Environment, nutrition, being factors, admixture with another race, domestication. Intermarriage spreads mental traits. Epistemology, the production of knowledge, was the key, an old philosophy by the Greeks, in the study of the nature of knowledge. That it was learned by preceding generation after generation, one's logic, the difference between primitive and civilized is like irrationality to rationality, one must work on habitual mental operations. And he could teach them. For they looked in need. It was at this juncture also, he no longer feared the long eared giant flora that seemed to guard this world, and gave him a strange feel on his pathways. And to be frank, they appeared to withdraw their gloomy look. Perhaps they knew this race of down and outs, actually had looked-out for them, wished to protect them; and was hoping McGee was their savior, to bring forth a newer and better, and more intellectual future.)

Thought Dr. Butch McGee, "Why did these women survive, I mean how?"

Always a scientist, he pondered this question long and hard. Did they have part of the Gerome of the Neanderthal like him? Really all humans did not come from the same ancestors, but yes, all belonged to one family, humanity, he concluded. So, they survived from the DNA of the Denisovans. It is hard said, but the truth is the truth, these women were survivors because of their immune system.

Now his question was to his second mind: could a billion years be interjected into a million, or better said, can one decade upgrade their primitive logic to a more advancement, to the 21st Century's way of thinking, or at least could they clutch onto its corners? That said, could he be the injecting tool of an anesthesia into their thinking, and reasoning? He would try.

McGee cohabitated with all seven women, and it came to mind in those early months, how quickly oxygen became so plentiful throughout the Dark Planet, becoming a part of this sphere, evidently only taking 1.5 million years, compared to Earth's atmosphere that perhaps took four billion years to form; again he knew, the dimensions in space were different, like at the beginning of the universe, its physics were not the same as it is today. As the universe expanded, so did its physics, its chemistry, its makeup, its learning, just like his seven wives. It was a mystery of mysteries, a quantum mystery per se.

He measured the air every few months, for the decade he was to remain on this orb: the molecules being O2 consequently he knew eukaryotic cells (being a single cell or multicellular organism) had good nuclei, and two sets of walls, a special protein bond and contact with the DNA, this protected the DNA, which protects the chromosomes. Therefore, whoever wanted the race renewed was willing to experiment with tribal interbreeding and those now in place, being: he and his wives and now his children, were but the bottom of the iceberg, the peak yet to be seen, and not seen for another generation. By and large, his children would have a strong immune system indeed, and an upgrade of intelligence of five centuries.

#4985/1-8 & 9-2016/revised 2-6-2016

Flung from Pluto

(Quantum, Ripples in Time)

I live in the present as most people do here on Earth if not everyone, which can be referred to as pragmatism, or logic, although I have experienced, as have my countrymen, what is theory to humans, and common practice to us: stemming from the present into the past, as well as into and perhaps the future. I am stirred for the most part by the here and now, what I'm accustomed to, even though I have come to realize the present and the future are well-appointed by God. Let me explain where and what I am leading up to:

Through a ripple of time, a ripple in space and time, nearly 1500-years ago, our spacecraft landed on Earth. We travelled 15.5 billion km from Earth's sun, in a matter of hours, not at its likely occurrence of several years. It was the year 488 A.D., when we came.

As I meditate and think back, the year now being 2015 A.D., the Picts from upper Scotland invaded Britain, as did in time the times of the Angles from Sweden and Denmark, and the Saxons from East Germany. Back then the land mass wasn't divided as it is today, this is to my best recollection.

Back then I still recall the limping, the hungry, the thirsty, the grim, the dusty and the sentimentality of the age, —notwithstanding, the horror of Warlike men from the North, now called Vikings in and around 790 A.D., and then after 828 A.D., when England became united, thus they were much like the ISIS are today. Both Godless, swollen with the devils' breath, bands inhabiting Earth, with no rules, indifferent to their fellow men, no limits, or discipline. Which leads me to my next thought:

I am a prudent being, I'm also impetuous, and our mission was to investigate the ripples in time and space. But as you may have already figured out I was left behind, and once left behind, there was no escaping Earth's gravity, its several layers of atmosphere. We are a long-lived race that had come from a planet half the size of Pluto, as I say in nearly the speed of light. All the same, I have always yearned to be reunited—yet on the other hand excluded this possibility of being reunited to the underworld of our little planet, although I've just read a scientist has discovered V774104, this is my world, a tiny

34

oval dwarf planet, Pluto minor, it could be called and would stand up to that name.

I made my way hobbling through the centuries, as if they were a descending staircase. But how this all came about was: one evening I had fallen to sleep, after having several horn-cups of sherry, when I awoke, it was a mid-summer's nightmare. I looked around in vain for my spouting friends, I had come with. And they were glowing in a heap of fire, as if they were the fuel. In a word, those who did not escape on our spacecraft, were roasted alive, and I had but a moment before I'd be lain outstretched before those cannibalistic mongrelized primitives.

Since then I have learned to look like humans, disguised myself somewhat. Now I am old, about to die of old age, and I shall explain one of the mysteries our race discovered, or think we have, we can put it into the basket of Quantum Theories, if not plain physics, we have learned everything that has taken place, in the Universe—as I say—and we can predict what hasn't taken place, to a certain degree, that has a high probability of happening, it is all written down in the ripples of space (and I say to a certain degree, because God lives in many dimensions, and who is to say, we even know half of them...) should you be able to scrutinize one you would be able to step back to let's say: 790 A.D., or to the end of time, if possible. The end of time is simply a mathematical equation in light-years, which stretches from one boarder to another. If we are to stay in our time, or Earth time, then my time this year is: 2015. And in a week it will be 2016. But at the end and beginning of God's light-years, it was quite different.

And so, here I am stuck, as I see in the news, war upon war, all leading to WWIII, as the ripple effect takes place:

nation upon nation taking sides with whom they feel will be to their advantage ... joining one side or the other that has now involved several groups of nations. You see, the universe is comprehensible because and only because it is governed by scientific laws, those God put down. That is to say, its behavior reacts accordingly, should it not, God forbid! And God has set down rules for man, but we cannot describe him in any mathematical language as we can gravity. He is a magnetic body that attracts the most incomprehensible actions that even God might say, "It never occurred to me...!"

#4915/11-20-2015 (Revised, 12-26-2015)

Voyage to the 10ᵀᴴ Planet

(With Dr. G.B. McGee/and Dr. D.) Part IV

In earlier times one would not be able to explain this adjustable journey with any credibility, and to be upfront it has all been kept secret over a half century-plus, upon this reading. Yes, to this date, 2121, A.D., it has been kept in virtual silence, but it took place in the year, 2050 to 2053 A.D., we can call it an old Mystery, renewed: of the 10ᵗʰ Planet, to which it exists in its own personal solar system, with its Paleolithic inhabitants (Neanderthals and Denisovans type, with evolutionary genetic immune genes, perhaps cousins to earthlings). This story you are about to read, has been taken out of the nation's archives, where it was kept for posterity's sake, for lost causes, until now. I am the Great, Great Grandson of Dr. D., whom worked at the once Louisiana Space Center, with Dr. McGee, and Dr. Hightower. And this is his story.

Dr. D's Journal

"On the 10th Planet within its own solar system its inhabitants live like wild beasts, if not often more so; before they die of a dark and lonely death in some hidden corner of its wooded areas, or volcanic surrounds (which tectonic plates seem to be everywhere under the planet's crust, and cause rattling quacks daily), thus many live in defensive packs, like wolves, most near strangers not long before they join a pack. They pledge their allegiance to one another on blood battles, the blood in the veins of another new comrade. There is no such emotion of bereavement found in them, although may I say, it is an emotion more learned than felt, learned by social comparison for the most part—if not a primitive impulse unlearned to this day, as well as an unlearned attention span, or having a high power of reason, in view of contemporary man, as having a weaker sustainability— thus, imitating behavior from the previous generation, and to give to the next to come, I will explain this further if need be; so, it was always there, it just was never brought out.

"It is for them a normal practice to torture unknown persons of other tribes, shoot a rock at their heads, club them to unconsciousness, and bury them half-alive under a foot of soil, in the bush. These things were learned by three means, in particular by my trip to the planet, and upon bringing back Samuel, as I will explain later, with the other two resources.

"So, it was wise that one does not lose contact with everyone else of their tribe, or wander off on his or her own, for survival purposes."

Behind a door in space (a portal, as it is often referred to) my great, great grandfather found an adjoining room between the solar system of this so called 10th Planet and Earth's solar system—figuratively speaking, like a wormhole. At that time the Louisiana Space Station had been monitoring muffled sounds of those ancient people, oh, just snatches of sentences and finally faint images. With a decade of this monitoring 2040 to 2050 A.D., the door had slowly opened and as Dr. D. explains in his dissertation in his third Ph.D., in Stellar Ages, "Now we can hear spirited conversations and even nightly fireside debates. And although we can take part in the exchanges, I don't wish to steer their thoughts one way or the other, lest we cause a collision which might alter their speed of growth and activity, and perhaps cause panic. This is all done by gravitational waves propagated, consequently we gain knowledge of the events on the 10th Planet, they are carried at the speed of light, but there is a new theory called 'Standing Still Time,' when this information seeps through the gateway, thus we get the near, moment by moment information and prior to the event information."

The account goes on to say: "What I wanted from the committee was to rescue one of the victims. And my friend, Dr. McGee commented 'Are they not an alienated historical tribe of nomads?' What I want—I told them— was to bring one being, or species of that race back to Earth to appease an embittered history they now had, and restructure his thinking, his reasoning, and if possible, plant one of his offspring back into that habitat and see if we can advance it. I told the committee, and I got smiles lit up on all their faces in the thinking room, especially on one of the indefatigable debaters on the subject (knowing good and well I had done a similar thing prior on the Dark Planet, which took a decade). By and large, I now had stopped them all in their tracks, to this grave question. Was it doable? Dr. G.B. McGee agreed, with me it was, 'Good luck brother! He said, 'Perhaps we may see in the long run the horrendous acts of violence stopped on that huge massive planet, those unspeakable acts of horror against its own population...'"

. . . Continuation

From the Journal of Dr. D., 2052, A.D., (now being read 2121 A.D. by his Great, Great Grandson, Antonio):
"The process of sequential metamorphosis for the body to change as need be from one environment to another was experimental, a drama in the process, of making our subject who came from the 10th Planet, ten-times the size of Earth, having discovered the planet 35-years prior, up to then it was covered by a next to invisible fog in space, —I repeat we were trying to have him modified, but were in fear of depression creeping in.

"Its orbit around its sun takes a million years we figured. The planet itself is a trillion miles away, its orbit being one-hundred and fifty times that of Pluto. I've named the planet: 'Mass' for its colossus size. How it formed is still a mystery."

"Well, we rescued a victim, of course not the one I had mentioned to the committee, and brought him back to Earth, named him Ssam, or Samuel for short.

"Now I must explain gravity before I get into singularity, to describe how this all came about, for those readers

40

who already know this, it will be a revision:

"Should one put a boulder or heavy brick on top of a straight mattress, the mattress will indent, depress, sink, there will be a curvature, should you put acorns on a straight mattress during this process prior to that indentation, the acorns will have produced a foundation to slide downward alongside the boulder. Now we know the boulder has already shape-changed the mattress, thus depending on the size of the boulder, and the curvature will decide where the acorns will settle, this is gravity. But we will take it one step further in a moment.

"The 10th Planet and its sun has such a curvature, which creates the planet's orbit. 'How so?' The planet's sun is collapsing, it's a gravitational collapse, and it is collapsing under its own internal gravity, the loss of energy. In other words, it's a tired elephant. When the sun does collapse, it will be devoured by a black hole should it get too close to one—to its Event Horizon, or just an inch over its assault-side? You see it is an exhausted star, all its nuclear fuel will contract like a woman in labor, sooner than later, and/or become a dwarf star or neutron star should it avoid the black hole, or—I repeat—food for the black hole, which is a furnace in essence, whenever.

"So, we seized the moment, and grabbed Samuel, should that event take place in my lifetime, and of course it didn't."

"When we landed on Mass, it was like stygian muck, a perilous planet with volcano gases being emitted every-which-way, eruptions all about, bubbles of gas, in many locations formed by the underground magma, it was a primitive earth in essence; hence, as we stumbled about in its thin and effervescent atmosphere, we noticed many strange animals the Dubbed Pegomastax africanus or

41

'thick jawed' mammoth rat! Old skulls of the Homo habilis type, stone tools, like Oldowan tools. Some of the tribes we encountered were likened to Homo heidelbergensis, often referred to as predecessor species to the Homo sapiens. Another thing we noticed is that they were capable of ritual behavior. And ate from its oceans the animal called Xenoturbella, a long word for a short creature that looks like a deflated balloon, has no eyes, brain, teeth, and is perhaps an ancient cousin to the bizarre looking jellyfish. These creatures are also found on Earth deep in the Pacific Ocean.

"At times it got so hot on this planet and at my old age as it was, we got fever burnt, that is to say, infectious diseases on our skin, even with our suits especially made for this buoyant thermosphere. At night we underwent surging winds as if we were on top of Mount Everest, and fading nights of gray, to a hot morning. A few of the team members, of which there were seven of us scientists—not to include the spacecraft's team—experienced delirium, a lucid mind, at times nearly madness came and went, it was as if the planet was cursed, but once back inside the spacecraft, sanity was restored. Our stay was brief at best, which is a story within itself of which I shall spare the reader and get right to the facts.

"At night we encountered the dead inhabitation, the spirits, fiends you might call them, the residue of the souls of the dead inhabitants of the 10th Planet. Evidently they live on, within their foggy world, but appear only to be a nuisance, more so than a threat, for they do not cause havoc.

"But our team members all survived and Samuel became accustomed to Earth with our metamorphoses process on Earth: at which time we eliminated his goatish and/or piggish stench out of his skin.

"As I have specified, it is for posterity I write this analog, between them and us. This work of course will be hidden in the archives of the Government for seventy-five years I expect, if not more, or a little less depending on everyone's curiosity.

"I also made a few more discoveries before I was done with my work with Samuel (who may I say at this point, died of depression) a few years after he reached Earth, and became our prototype for future escapades; you see we did not let him live a life of his own, we scientists put him into a closed and guarded environment, accordingly he never had time to produce offspring, to circulate within our environment. One glorious discover I might conclude was—

"I discovered that he had the same genetic material of our earthly Paleolithic Ancestors, exquisitely preserved in his DNA, more than 20,000-years ago. My conclusion to that is that they had been brought to Earth, to do mining, perhaps in Peru for gold, by alien spaceships, realizing all spacecrafts need a certain amount of gold, it is a precious metal at times hard to acquire especially in interplanetary space travel, and much needed for certain parts of the vessel; that being my best and quickest guess. The other extraordinary finding was, bringing him to the Chauvet Caves, with those drawings on the cave walls of those beasts that roamed Europe, during, and after and before the last and prior Ice Ages, or some 33,000 B.C., he recognized them. And I found out he knew how to use the compounds in recreating them: as he did, out of charcoal and red ochre, he etched them into limestone, to the likes of those at Chauvet. He evidently at one time, had been an artist.

43

"Samuel was also familiar with the stalagmites and stalactites, as if these ancient people were brought back and forth for 15,000-years, then left to their own, or abandoned on their planet thereafter, for 20,000-more years; who's to say? Whatever the case, they did not advance.

"A last word to the reader, of this 22nd Century, I presume 22nd Century: when I did a cross-cultural study of humanity's DNA, we found 3% of it to be similar to the origin of Samuel's."

#5047/2-2-2016

UNLEASHED ASTEROID

((2019—EaRtH BOUND) (WItH DR. G.B. MCGEE))
PaRt V

At the Louisiana Space Station, Dr. G.B. McGee, has just released to the radio staff, and television networks, information concerning the bombardment and strike of a runaway Asteroid that hit the Gobi Desert today. What had taken place—prior to this event—was that a supernova (a small star/sun) explosion took place in one of the one-hundred-galaxies that can be seen behind the Milky Way Galaxy, explicitly causing the creation of a gravitational wave (GW), which has been traveling at the speed of light for some time now.

Dr. McGee has speculated this wave from what is called electromagnetic radiation, whereas he has determined it to be a gravitational wave on the grounds it does not interact with matter. It has traveled throughout the universe from unknowable and countless galaxies, unimpeded, silently, and has caused an asteroid to wobble off its course, the size being 0.6–miles wide and it has hit ground in the Gobi Desert. Had it been a six-mile wide asteroid, the size of the one that wiped out all the dinosaurs eons ago (the largest that has thus far hit Earth, killing 90% of marine life, 242-million B.C) well it would be a different story, —So, Dr. McGee has inferred to his audience. Also, he has told the radio audience as well as his television audience, in a matter of fact way: had it been the larger one as he described, he wouldn't be broadcasting the Asteroid malefic destruction at all, plus there would be such a climatic impact not only on land but atmospheric worldwide, we'd be back in the Stone Age. In addition, he has also openly told the radio listeners: had it hit the ocean, it would have been even less severe.

Dr. McGee, goes on to explain in more depth: the cause of the movement of the asteroid: "Perhaps an eerie mystery to many of our listeners today," he comments, then goes in detail: "GW's are a ripples in space, waves that are concentric ripples that squeeze and stretch the fabric of space-time, and are caused by the movement of mass, as in the case of a supernova explosion (a dying sun). So, the GW passes through space, as it does, it squeezes and stretches that space it passes through; they can disturb and upset what is in its pathway. A GW can also be a window to figuring out the origin of the Universe. If indeed one could see and read a GW he might find himself a second away from the Big Bang. But now to more serious matters."

Dr. McGee, goes on to explain, the happening: "The crater, in the Gobi Desert is some nine miles wide and it has thrown huge amounts of dust into the atmosphere, and because it landed in the desert, it did not trigger worldwide calamity, such as: large-scale fires, —thank God for that! Although the soot thrown into the air will circle the globe restlessly for a while. Expect the atmosphere to remain with soot and dust for some ten years, worst case scenario. Hypothetical, six years at best. Also expect the particles will warm in the sun, thus, heating the stratosphere, destroying, or at least speeding up the destruction of the ozone, and expect harmful ultraviolet radiation, worldwide."

Dr. McGee has told the people not to panic, but to be realistic.

"In the days to come," infers McGee, "70% of Earth's surface, its sunlight will be reduced for one to two years. It will cool to 14.5 degrees Fahrenheit, which is a new, but little ice age for the world to deal with.

"Thus, global cooling will drop its precipitation to perhaps 50%, this will of course cause damage to plant productivity, especially in North America, Europe, Northern Asia, India, but South America and Africa should not be affected as badly."

The station manager has told Dr. McGee, not to be so bleak, to try and give the people hope.

"This space rock, which was more like a small hill, or mountain, is one of some 900-asteroids that pass near the earth regularly, and at the Gravitational-Wave Observatory, scientists using the Laser Interferometer, do not predict any more GW's of this sort, or Asteroids to be knocked off balance. NASA officials has indicated this mountain-size space rock was simply passing through Earth's neighborhood at the wrong time, had we known, had our president been alerted to this sooner, we could have perhaps avoided—that is to say, had our president known the object was about to hit Earth, he could have used a nuclear bomb, destroying or disrupting it, giving it a nudge to the right or left a tinge, as to slam the space rock out of Earth's path. But by the time it got to the officials and the big shots to make up their minds on what to do and not to do, that option was out, thus we had to simply sit and wait, and this is what we get for waiting, when you snooze, you lose. It is one of those unseen and un-seeable events that perhaps could have been imagined and foreseen as far as, avoiding this climatic event, had the right people listened to the astrophysical specialists."

#5064 (2-2016)

48

McGee's Colossal
Shockwave
(With Dr. G.B. McGee) Part VI

Eatable Water Creatures from the Black Planet

Knowing all explosions fade, even monster explosions, billions of light years away, supernovas that is, perhaps at its surface being 100,000 degrees, this mega-supernova with the brightness of 600-billion times that of Earth's sun, 10,000-light years away, Dr. McGee standing on the Black Planet looking up at it and into the Spiral Galaxy's far-off darkness, into the constellation #89, knew its center to be very compact, boosted by a dense highly magnetized, magnet. All of a sudden a shockwave disturbed the space environment around the Black Planet, —causing McGee and his family to find shelter underground, thus the atmosphere filled up with stirring gas and dust, blinding the planet from its sun and moon, dropping the temperature to a near freezing level for several months, killing much of the foliage and all the long ear plants that had evolved in the last million or so years.

Actually what took place—Dr. McGee explained to me one afternoon, after his return to Earth from the Black Planet, and I shall explain what he said now, the best I can: When this explosion took place there was a Per tantrum (a perturbation), throwing off enough heat and radiation to make today's sunspots and coronal mass ejections look like hiccups. This radiation vaporized most of the Dark Planet's crust, the solar wind had blown the remnants back out into the solar system around our planet. Saving perhaps a good portion of the foliage, yet not nearly half of what it was. But for less than fifty-people, it would do. Had the Dark Planet been a larger planet it might have lost its outer shell, and become like Mercury, a dead planet, although Mercury is small, the Dark Planet is smaller. We didn't have big collisions either during this cosmic shockwave, commonplace on many planets and moons as you know Dr. D., knocking us out of our orbit, —it could have done that, thank God that it didn't, we'd have plunged into the sun, and I mean the nearest star. But that's how it was, if that makes any sense. Actually what I remember, to be frank, was a swirling disk of dust and gas, like a solar nebula, circling the planet, what stuff or material I can't say, but blinding, hot and scorching.

With McGee's several wives in a warm marble like cave, with grand peaks and towers, carved to the likes of a palace by nature, and with his twenty-one children life went on nearly as normal; a creek ran throughout the cave's maze, allowed for a good water supply, and odd looking fish and other growths that were eatable, hence, it was under such circumstances that for months the inhabitants of the Dark Planet lived in a semi state of hibernation; and thereafter, for Dr. McGee, things returned to sub-normal.

But all is not well, that ends well, he was getting tired of living a hermit's life on the Dark Planet, even with many wives, for his heart was on exploration into the cosmos, not stagnation on a planet.

#4996/1-14-2016/Reedited 9-2016

SWING PLANET

((WITH DR. G.B. MCGEE) (2040 A.D.)) PART VII

The Louisiana Space Station had discovered the 9th Planet's exact location, tracing its orbit for four-years, on its 15,000-year orbit around Earth's sun. Being ten-times the size of Earth, and a thousand telescopes on Earth searching for the new planet alongside its orbit, and Earth's new outpost on Mars doing likewise, it was ambiguous, but just a matter of time.

Compared to the Dark Planet, it was a hope-skip-and-jump to its rock forest, called the Kuiper Belt, where its orbit centralized, and passed Sedna, a planet nearly the size of Pluto hidden within its channels, and when it did pass Sedna its distinct alignment with the rest of the solar system was mapped out to the least fraction, but only within that area of space.

In the adjacent solar system, objects unnamed, star watchers found the missing planet as it zoomed by, covered by massive debris of asteroids and alike, thus not very visible, and as it rotated on its spin, its axis, objects swung outward to and fro around the planet, as if protecting it from bombardment, and the large planet had a pull, a dynamic force on its pilgrims as Dr. McGee coined them to be, on these far-flung orbiting moons of sorts, which seemingly stayed in a permanent stage of circling its mass, but also they shed light on the planet, as would a comet. And the closer to Earth's sun, the more energized these objects became, and a brighter light did they shed. And the more visible became the Ninth Planet, as it was called by the media, or X-planet, which was really out of date, something out of the Flash Gordon days.

Dr. McGee, as usual had ideas, like the one he had for the Dark Planet, if only he could find a way to harness something on the 9th Planet, to explore it on its ancient journey, for mankind, the happier he'd be. The superintendent, Dr. Hightower was of course impressed with McGee's idea, but as he told me, during our last meeting—on the side—and I shall express it the best I can:

"We're still trying to figure out the birth of the Spiral Galaxy, it would seem it was formed like the Milky Way, or started in that manner, from the middle and grew outward, and as small as the Spiral Galaxy is now, in a million years, it will be ten-times its now size; so, I concluded with Dr. McGee."

Said Dr. Hightower:

"Stellar Spectra age is difficult to get, even consider, I wish Dr. McGee would stick with that. This new Galaxy he discovered #79, is starting out as a small disc, and in time will grow outward from the inside out. This we might be able to see in our own lifetime, through the aperture, we've discovered."

When McGee saw the black hole before it exploded and hiccupped its insides, we got to see a crescent shape as it swirled around the black hole.

"We need you, Dr. D. to keep him on track, if only you could!" expressed Dr. Hightower.

#5021/1-21 & 24-2016

Far Boundaries of Gravity

(With, Silas Caine)

The planets, move around the Sun, and the Sun with all its planets and moons, thereof, revolve forevermore, or so it appears on their straight and narrow orbits, as the universe pours out with care its widespread gravity, like manna.

Gravity, the thing that keeps our solar system activated, our galaxy in motion. In essence, the universe's black energy, black matter is the universe in a nutshell, take that out of the equation and everything will fall on God's lap.

Should a star collapse, we know it's an act of gravity, to a certain extent, as then a black hole is ready to be created if not already created, to swallow that dead star, whole. Yes, the hand of providence has provided a waste basket, for its surplus. Put another way, gravity shapes space and time: as energy, creates a reaction, produces a black hole, if not a wormhole that may develop into a black hole, or a black hole acting like a wormhole.

This must be explained before we go on with the story lest we get fragmented along the way, and we lose sight of Silas Caine who is the antagonist of the story.

We have talked of gravity, now wormholes, and travel, then the full story, since the title is "Far Boundaries..." it should fit. What is faster than the speed of light? Nothing, scientists say. But if you step into Quantum theories, physics, there is something perhaps! Traveling through a wormhole which I shall define shortly. To make a long theory short, a traversable wormhole, let's say with a notion of an intra-universe connection, the wormhole being similar to a black hole, more on the order of a shortcut connecting two separate points in space and time, you know, like a tunnel, with an electromagnetic field of energy. In essence, as Einstein has inferred: a wormhole bridges, in this story, two different universes; —we're getting into general relativity, —the wormhole might allow superluminal light speed, faster than light travel; the speed of light is 186,000-miles per second, this again is more into Quantum theories; meaning, hypothetical, unseen, unproven, or not experienced, but probable.

Now a third thought into this story before we get into

the belly account, and its tiers: since we have taken care of Gravity, and now Wormholes, which we may come back to—that are much like a hole in a cylinder, and if we can stretch our imagination we can envision it being used for communication between parallel universes. A side thought, wormholes cut 50% to 99.9% percent of travel time, or put another way, a billion miles can now be measured in feet, we are on the right path to knowing how Silas Caine got to Earth in the first place, from that far-off universe.

So, the next step into the story that I haven't yet told, is time-travel (see end notes), like a time machine. In wormholes, time is not the same as outside it. We are now going into Quantum effects if not general relativity: Black Holes are of course Stephen Hawking's expertise, and we have and shall mention them time and again throughout this account. But it makes no difference, what is, and what is theory, is perhaps possible, there are many realities we know nothing about. In this case, and a possible case indeed, and for the story, a particular case, the wormhole is a time machine, call it time dilation, or contraction. We are traveling from one universe to another, faster than the speed of light—even though the light in the wormhole will beat us to our destination, and let's say we are carrying energy from one time to another, from one place to another, the wormhole has to adjust, or once on the other side, something has to. Now we are ten-years in the past, or perhaps 1000-years in the past, or we are from the future visiting Earth in its present, we now are looking at time warps, or twists, or time like curves. All this and a little more, concerns Silas Caine's alterative and altercative activity that has been disputed vehemently!

Silas Caine

And so what we know of and what we don't know of, moves unless balance is held back by another force, but when someone can trigger a force against another, then what? When perturbation accrues, when a traveler or something physical, a system can be disturbed, even planets taken off their orbits, is when there is an interference; Silas Caine knew this, and used his Severity Machine, as it was referred to, he brought from another universe to do just that: why? To see if it would work, and then return to his own universe, via wormhole, and let his countrymen on his planet know of his experiment; what for? He wanted the Nobel Prize of that Planet, which was called, the 'Quantum Award.'

Two forces fighting one another, and thus comes balance or chaos. What we don't know God holds back from mankind as a father would hold back a grenade from a child's hands reaching out for it! Why? Because the child or we, are not ready for it, and man is quite predictable, for obvious reasons, but when an outsider sneaks into your solar system, and onto your planet, predictability changes, and thus indifference takes precedence for the invader. So, the question may arise: what does he know that we know, but live in pretense of not acknowledging to know? A riddle? Perhaps not. Extra-terrestrials are out there? For good or bad, who's to say, and what is on their agenda?

The breakdown of gravity is power. And so it was in the year 2036, there appeared such a man by the name of Silas Caine, who had built a machine—to repeat myself, an awful machine that revised the pull from the iron core of the earth, allowing Earth's atmosphere to be heated, polluted, scorched, and thinned, and there it faded into the empty space between Earth and its moon, like clouds.

Where did he come from, how did he build that machine? Some say he just appeared out of nowhere, others say, he came through a wormhole from another universe, that he came from the future and brought with him, his own energy from that far-off place. And now man devoid of weight, started to become like fiends, ghosts.

He was cloaked in human skin, a fabric he manufactured before he arrived on Earth, his actual appearance was as described: eerie, enormous open nostrils like a cow, with flaring ears, likened to an angry dog's barrel chested like a gorilla, lanky limbs like a monkey, rudimentary slits for eyes, a dripped like animal tongue, about five foot and eight inches tall.

Slowly he evaporated 50% of Earth's atmosphere, its global oxygen, carbon all the emanating elements that form the timid and mysterious invisible atmosphere, that took billions of years to form around Earth; consequently Earth now was in a state of loss, as was Earth's electromagnetic energy field.

In a short period of time the oceans were lowered to the point, ships could no longer navigate them, and the world powers became uneasy with one another thinking one another had some kind of conspiracy going on; and so Silas Caine, to Earth's once thick and luscious layers of atmosphere, nearly all was cast out into the black matter of interstellar space? And no one could find, or stop Silas Caine!

Earth was on its way to becoming another Mars.

As a consequence came, underground sanctuaries.

He could have controlled the world, if that was his

mission, he had the power, but that was not his mission, as we all know.

And before he departed Earth, he did leave something intact, Earth's culture. But let me backtrack to those latter days...

The underground vaults of Earth vomited up all it had. Billions of voices screamed, and ascended into a cosmic death. Satan's demons had a great feast, and long laugh over all this, they didn't even know who Silas Caine was, or where he came from, but was as if he was their Antichrist!

Did Silas Caine want to get even with God; so, the preachers and clergy summed up, when all other reasonable thought fell into a black hole? Logic would say he had no God, and didn't know of any.

"Alas!" "What now?" were the words being uttered daily in the newspapers commentaries, on the talk shows of television, on the radio programs. Was Silas Caine as they said, 'A lunatic?' How could they stop that machine, when they couldn't even find it? Then puff, he was gone, liken to the way he came, with that darn machine.

#4916/11-20-2015 (Reedited and revised slightly, 12-26-2015)

End Note: traveling at the speed of light (if you could) your aging would slow down immensely, and could be a form of time-travel. Time stretching (or dilation), in other words you could then step forward into time past, considering your travel is away from Earth. Why? Earth time is faster than your time (the net increase in the Earth's time relative to yours is -1 +4 -1 = +2.) Assuming you will come back to Earth, you have aged one year less going, and one year less returning. While for the most part, Earth is stationary.

The Mineralogist: Night Termination!

(With Dr. G. B. McGee) Part VIII

What makes Earth special?

The rule of war is to avoid hitting hospitals, the International Criminal Court calls it a War Crime to do so. Now there is a new war game going on, the rule of proportionality has been altered, which says, even if the enemy is in a hospital or medical clinic, you cannot kill the enemy, because the greater number will be civilian casualties, thus how do you kill the enemy then, when they are hiding in the very place that is Internationally forbidden to be bombarded, it is an enemy's sanctuary? The Mineralogist has come up with a new weapon of war, 'Tight War Termination Killers'. Prof George Bacon, Mineralogist and Entomologist (the study of insects), was assigned to the project called: 'Night Cessation' (or 'Night Termination'); for the United States Defense Department.

So, the question comes up again, what makes Earth special, more so than Mars, the Moon or Mercury, to mention but a few planets? Minerals! In a nutshell, minerals!

Recently cobaltarthur—5 mm were discovered, and we have a tinge of abelsonite, and fingerite, and particularly a thimble full of edoylerite, the latter being called by those who know the mineral category well, the Mineralogist, and more often than not, the Entomologist, who study insects, especially those species covered in amber, or fossil flowers trapped in chunks of amber for fifteen-million years. So, you take the Mineralogist and the Entomologist, and edoylerite, and name it 'the Vampire-like mineral' and what do you get, you get a mineral that decomposes on exposure to light (as rare as the universe is old), a very rare short-lived mineral, and hard to find.

Now if you take the mineral fingerite, it forms like flakes on the Izalco Volcano in El Salvador—the only place in the world, when it rains fingerite washes away, gone forever. Thus only a few people know where these special mineral reserves are in the world. Safer that way.

Rare as they are, can be manufactured by industry; so, Dr. McGee asked the Louisiana Space Agency, upon the visiting Mineralogist, "What insights does it provide for us here, and America as a whole?" Prof George Bacon, the Mineralogist, ventured to say—the three Star General, Wessington at his side—whom both were working hand in hand with the Underground Carbon Industry, "It all works in with minerals, Earth's system, insects, the colorless crystals are essential, they produce microbial 'poop.'"

Now they were all sitting in the meeting room and the General asked for them to turn off the lights, to make the room halfway dark, and Professor Hightower did as asked, and someone walked into the room, one could hardly make him out, "This is our new night fighting machine, a soldier of gallantry, like the Green Knight!" said the general.

Dr. D., looking closer at the outline of his face, a mixture of reds and browns and greens, had the room been any darker the person would have been un-seeable.

"What exactly is he?" asked Professor Hightower.

A little light came through the curtain of the window, and the General and the Professor hesitated to answer, as the soldier of his genera, stepped back under a shadow, as not to be inclined to have the light directly upon him. Then Dr. D., took it upon himself, since the General and the Professor were silent, by not answering the question, pulling out of his pants pocket a small Boy Scout flashlight, attached to a compass, and shinned it towards the object direct, or night soldier, as he was being referred to. Instantly he became decomposed, moldy, rotting away as those flacks called fingerite, and not so unlike the mineral it was made out of edoylerite called the Vampire Mineral, and there he evaporated, the longer he was exposed to direct light, the more he become nothing but residue.

"You did exactly what we predicted you would do if you could figure out what we did," said the general, "and now you can't question our hunt saboteur, or assassin, once inside those forbidden walls that have International Safety attached to them; whatever mission he may have, the enemy will never know. He is an enemy's nightmare."

#5074/2-13 thru 16-2016/reedited 9-2016

65

Charon

(The Listeners)

The explosion had taken place outside the Kuiper Belt, those being on Pluto's moon, called Charon, one of five, had heard it, as did the Great Scientist Alanxdro Gessen from Earth; thus in his lab the green light went on indicating, predominantly the strange and wonderful had taken place, an opening of communication between Charon and Earth.

Perhaps a neutron star very dense burnt-out, or the merging of black holes took place, but all were guesses; yet Louisiana's Space Center was listening as was Dr. Alanxdro who was doing likewise listening in his lab.

This happening created a gravitational waves (GW), which created an opening for Dr. Alanxdro to be heard by Charon; hence, the warping of space and time took place, he conjured, what may be called an astrophysical phenomena.

The explosion, perhaps a supernova, took place some place in the Milky Way Galaxy, he concluded, with the assistance of the Space Station.

The GW, a ripple in the fabric of space and time, came at the speed of light, at which time the Louisiana's Space Center was talking and being heard by whomever was on the other side, on Charon, as was the old scientist. It was as if there had been a tunnel, or gateway had been created for Earth could hear the voice from Charon, or vice versa.

It was the first time mankind had heard another voice that belonged to an alien race, some 2.6-billion miles away.

Then after a few utterances, some form of matter impinged both the earthling and the alien radio waves; both turning their knobs on their wireless transmitters getting only inharmonious sounds each to the other, both trying to figure out their meanings, but it was impossible to interpret them now: nonetheless contact was made, and whatever created this opening, it now was closed.

Written: 11-21-2015 (4:00 a.m.) #4917

Enceladus' Dilemma

((The Account) (Year: 2035 A.D.))

I glanced through the new telescope, the largest in the world, I was told. Far did I see in the depths of space; for hours I searched our solar system, focused toward Pluto, and beyond, some five billion-km plus. The telescope was incredible, connected to another telescope some 200-miles above Earth. Somehow I found myself fixed on Enceladus, Saturn's moon. (—First let me say this before I continue, point of fact: physics alone cannot explain reality, if it could it would explain God, and since it can't and can't take God out of the equation, without having it collapse, they have to conclude there are alternate realities—with this in mind, I proceed with 'the account'.)

So, what did I see exactly with this newfangled telescope?

A man in a boat, or on some kind of platform, on ice or stilled water. I glanced at him swift, surprised. He was the size of a sparrow. At second glance, I adjusted to enlarge in size, a typical sort of thought came—: nothing in the astronomy books on what to do on delusions versus reality; hence, was this a delusion, or reality? I was overtired, several hours had passed by; surely the psychologist would adhere to this as a delusional syndrome before reality. At a third glace the item became sharper with more magnified details. Now he looked more like a giant size bird, let's say vulture, with shark details of a sort! Then with adjusting the focus, it became a man with a head and wings, and legs, or could they be some kind of wing attachment, or oxygen tanks? I asked myself: how does one rationalize this, then I looked for the camera equipment, and could not find the panel, which comes under: lack of instructions.

Enceladus I knew was a rich moon, being an amateur astrologer, my dear friend Hkanat, —who allowed me into his astronomical temple, had only expected me to observe, like one would at a museum, or a fish aquarium, and had to leave me on my own pending other duties. But as I was about to say, Enceladus is a most promised place to find life, if life is to be found anyplace in our solar system, in particular beneath its icy crust. Should this moon, warm up, it would be one large ocean I have confidence in.

70

Consequently what was I seeing? I mean, really seeing? I knew in advance there had been found years ago, molecular hydrogen, also hot vents on the rocky ocean floor, this all proved habitability; perhaps like Earth's deep-sea beds. Thus, Enceladus like Earth—was both like to like—in this manner. In all speculation, until now, no one could prove this moon could support life. Not knowing all that much about the Saturnine system, I could see in the wink of an eye, this underwater wrinkled terrain, its mountaintops soaring above its atmosphere, may indeed have had done just that, supported extraterrestrial beings at one time, or otherworldly beings from other planets, as not being a delusion here was one of the beings. Mentally my puzzle was being put together rapidly. Perhaps a portion of this moon had been, or was, in a process of serpentinization. With more adjustments I could see rocks rich in iron; my awareness of Earth's micro-organisms—through which this same process was a source to drive Earth's metabolism, like protein drives a man's metabolism, why not here on Enceladus, like the sun does for a reptile?

And then it happened, at my next glance, I felt helpless, dread, which in the years yet to come would I dwell upon immensely. Did I depict an overlapping of time? In other words, did I survey time past as in a second print? A silhouette in the ripple of time past? That was still outlined, definitely in the mind of the universe? The question begs to be answered: why did not anyone else see what I saw, especially my amigo? Disregarding the present. As if in death, man is said to be able to reflect all his past events, how is this possible? Again I must say, like to like, God's universe has preserved it somehow. In any case, when I explained this to my colleague, he said:

"Different realities are possible, in cosmic time," and he left it at that. I took his statement to mean: what is fretfully unexplainable doesn't mean there's not a reality to fit it.

Written: 11-22-2015 (#4922) reedited, 12-24-2015

ARCH DEVIL BELPHEGOR'S
DIAMOND PLANET

Copernicus, also known as Planet 55 Cancri e, or the Diamond Planet, twice the size of Earth, some forty-light years away, that has an orbit so close to the sun, its surface is 2000 °C. The planet takes eighteen hours to circle its sun. Its mass, eight times that of Earth, and it is rich in carbon, inferring rich in graphite, coal, limestone and petroleum and above all, diamonds, all commercially important fragments for Earth. It is said, the smallest diamond is the size of a large thumb, and the largest—of which are many—the size of a five to six foot stalagmites. The problem for Donald Rump Olton, billionaire industrialist, is that its atmosphere is much too condensed with hydrogen cyanide and prussic acid, thus poisonous, and no water, and very hot. Hence, one can see with the clap of an eye, this planet is for the most part, sealed off to any earthly invasion. Or is it?

As the old saying goes: if there's a will, there's a way! And Mr. Donald Rump Olton, he was going to find it one way or another. From Earth, it looked no more than a common pebble, a closer look, it looked like a chicken egg. Then Mr. Olton's scientists discovered a crack, a large fissure in the larger end of the planet, like the Grand Canyon, and as deep as the Colca Canyon in Peru.

For the industrialist, Mr. Olton of New York City, diamonds began to run gradually inside his head, wheels made out of diamonds, he had to perceive, recognize the unperceivable, and create a mechanism to mine them. There came that industrial devil's whisper of a hundred flutes in his head: faster and faster the diamond wheels revolved. Then came a thought, 'Did not King Solomon put giant demon to work for him in building his temple?' Yes indeed so legend says he did. "Why can't I?"

Thus, he called to the Master Demon, Arch Devil Belphegor, who dominated the netherworld. When he appeared there was a terrific suction, an unseen current more powerful than a 6.5 earthquake, had there not been gravity and a thick atmosphere above Olton's head, he would have been plunged into black frigid space.

As well as I can recollect there was no loss of time in the meeting and at its end, Mr. Olton had signed his soul to the devil's dominion: "Nothing for nothing" said Belphegor. And hence, the forces of the netherworld had a job for twenty-years to look forward to.

Belphegor selected a force of twenty giant demon, mostly complainers, complaining while on the job of numbness, painful tinglings, blisters and rosiness from the 2000 Celsius surface mines on planet 55, Cancri e, but in time they got accustomed to it. The good news was, the demonic labor force needed no water, or pure air, and therefore steady and slowly they could plunge through the sun's flames as the planet circled its star, somehow these unfamiliar spirits became analogous to the sun in a short period of time, adapting to its unpleasantness quite well, and not so unlike the Earth's molten under crust of magma, and mostly unaffected by its heat, just a tinge irritable with the fluttering of the sun's fire that burned into the planet's surface, more often than not.

The whole world could not guess where Donald Olton come up with such large diamonds, and until his death the demonic force throbbed with unnamable energies to dig them out for Mr. Olton, to build his towers and skyscrapers, the world over. But all good things come to an end, consequently, on his deathbed, he lay with an alien sadness, it was all so short lived.

Written: 2-17-2016/5073

METALLIC—BODIED BEINGS

(2045 A.D.) ((LETTER TO A DEAR FRIEND)
(A METEORIC PHENOMENA))

"The year is 2045 A.D., and man has come to the point that he can, and does qualify as tenements of the new so called metallic-bodied beings, their brains incased in a spherical domicile; beings within shells who now can live a thousand years, so they are told, but that of course is still theory, yet to be seen.

77

"Their eyes kaleidoscopically, are made up of some kind of: constantly color changing recherché adamantine material; when the eyes blink, they emit flashes of lightening, bright as the North Star. Their voices are clarion, and still maintain—for whatever reasons—human respiration, although it is an add-on, and not so perspicacious in cost. The globular heads are triangular, the metal beings have a sort of cupola back, and it is very hard for genuflections in the church I am told. The human beings that choose not to buy one of these lasting devises, calling them marionettes-freaks, with brains, are being considered prejudice, not sure if I can agree with that but who's to say. They are not arabesque. Nor do they need to eat like humans, although they need sleep for the brain to function, and a flow of nutrients likewise. Most folks who have purchased one in their old age, have kept their old bodies in storage, as in an urn, some made them into statues of case-hardened ash. There is a negative to this I found out, they are subject to the ravages of some corroding acids, should someone take advantage and pour this acid on the metallic-bodied beings, in sleep, of which they only need four-hours per twenty-four.

"The amazing thing about this new form of being, he has super x-ray vision, diaphanous. And people are complaining of this, especially in the bathrooms around the city here in St. Paul, Minnesota. The Democrats find nothing abysmal about this, although the Republicans do. There's no need for gun control, bullets will not hurt these metallic bodied beings; so, the issue of gun control is neither here nor there. Inside their torsos, or upper body parts, are frames of spiral rods and arabesque filaments, quintillions; a masterdom of science technology. Some folks who care to be different have even ordered the shells to incorporate a put-on and take-off proboscis, like a trunk for a nose; some have even ordered artificial wings to attach to the back of their metallic shells. I have ordered one myself, being at that age of enlightenment, to avoid the sepulchral, although I think I'll save the body, I might as well as for memories of how it once was. There has been some dirty dealings with this as often people will take advantage of modern science and its gifts: those with a weird prerogative for the most part: one doctor, I can't say his name for legal reasons (I'm afraid of being prosecuted, as everyone is today), has crafted one of these metallic bodied beings, and took it a step further, made him into a anthropomorphic, that being: half-canine and wholly diabolic, in that he now craves human flesh, a brain eater; I know you'll ask, 'how so...' by implanting old genetic material from a Neanderthal's bone marrow (spinal marrow from his spinal cord, the choicest, from the pith, to give vitality), into his brain. I told my pupils at the university, where I teach, 'Nothing is perfect,' like our president has inferred, and one has to expect some chaos, it's common, especially during the adjustment period of new ideas, it's a simple matter of new criterions; change is inevitable, change or be hanged...that's the name of the game, implying I think, it's even healthy: like Robert Frost inferred, so long ago: 'I like a little corruption myself, if it's amusing.' Incidentally, with these new pewter-like bodies, for a premium you can get two sets of eyes, kind of comparable to a spare set of glasses; although I hear the second set is dull and lifeless. On another note, I imagine it will be a weird ordeal, and the men whom I've talked to have felt as if they were being dissected. And I also heard: after the transfiguration, one's voice and language tone becomes somewhat similar to: horn-like intonations for several months, a 'recuperate period,' so one has to expect that.

"My operation will be done in the diurnal period, nighttime is too spooky for me. I look forward to linguistic studying thereafter, when I can learn two or three languages at one time, evidently there is a magnetic force within the shell that helps the brain in multiform wonders such as linguistics, my wife keeps telling me as do her friends I should learn better Spanish, well here comes the chance. But the best asset yet in these metallic people, they've found, is that one is exempt from all the ordinary biological needs and desires. That is some form of pre-metallic stage I've got to see. When this process is complete I can devote my time wholly to reading and writing and research. Although my wife says, I have already done that for way too long, and the infinite grotesqueries which I've devised and created with them, are enough for anyone's lifetime. Some of the side effects I understand can be: anti-social impulses and actions. As I stated before: nothing is perfect: ere a means of retardation as well. Heretofore, all the experimentalists have made that per near a doubtful reaction. The reason being, even if the body and brain are blasted into a million pieces, or fragments, they now have the knowledge to baste the brain back together—figuratively speaking—the body is useless anyhow. Yes indeed, they can reunite single atoms, electrons and protons; so, why worry! All one has to be concern about is the onslaughts of those brain eaters I talked about, and how many will end up being of that caliber? One out of a million. Well, dear friend, I got to go, see you in a year or so, in my new metallic body."

#5300/7-9-2016

A Short Vignette on Homo Erectus

(In Poetic Prose)

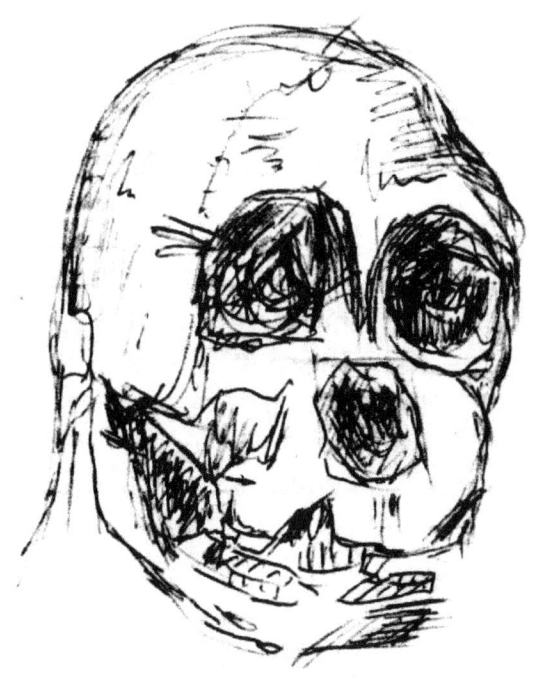

The terrestrial beast-man, some call Homo erectus, a million years in contemporary man's past—
Think not that God lost sight of him.
Perhaps God stimulated the atmosphere, thus, the unintelligent man-beast acquired a richer mentality, intelligence, now a creature of reason.

One may even call this, accelerated evolution—
Long before God created the Garden of Eden.
Why? For humanity's sake; for man to come out of the fog, and create his own destiny.
Perchance micro-organisms in the atmosphere, made him, erect (upright), and liken to a reversed plague.
Through a narrow process of metamorphoses, man was lifted out of the stage of a hyena-like existence—
Although avidity, along with: power, pride, and greed, man fell back into his old impulsive, and repulsion, once lifted!

Now man proved to be again unworthy—

Not so unlike modern man!

And thus, God sent them, the old ones and the not so old ones, one by one to the River of the Dead, where he assigned Charon, the oarsman, to ferry them out of his sight, to a land called Hades, where others have called it Tartarus, and still others, Sheol: where the dead dwell in darkness, and where it would be, for God, out of sight and out of mind.

#5348/8-29-2016

Discovery of Planet Moiromma

(And its inhabitants' biological system; by Dr. B. McGee)

A resemblance of Nori Iron

Dr. McGee explained at the Louisiana Space Agency, after coming in contact with a diary from a being from the Planet Moiromma, outside of Earth's solar system, and

83

learning of those he killed in his fifty-year spree in the jungles of Guatemala; and according to the inhabitants of that local, he had a god-awful looking presence. This for McGee was proof aliens have been coming to Earth for centuries; yet one marvelous thing about this discovery was the biological system, its chemistry of the alien life, and now he had a diagram from the diary and notes, and ashes found by this being's disintegrated body; in essence, he has formed this thesis from four particulars: empirical data by others, chemistry, the diary, and visual pictures of the far-off planet from the Space Agency. He is now trying to explain to the community—at hand, his findings, they are in a staff meeting:

"I've learned at first hand death to the inhabitants of the Planet Moiromma which we've discovered just outside our solar system, their early deaths, are like imprints, a carbon copy (or a facsimile), one after the other, they have many. Having said that let me get into more details. Sub molecules of dust within the universe dissolves into the spiritual matter of their souls their genetic makeup, and upon impact when they arrive on another planet after a, would be death, it is simply that a storm inside of them develops, seeking an exodus, or migration, and thus becoming less dense, the figure materializes. Hence out of this mass of vapor and dust a solid form is reproduced from the imprint of the depth, a new product is created of this same being. So in spite of the awful fate that might have befallen the Moirommalit, consequently one out of many deaths they can endure, disintegration is not the end.

"One must remember the soul is not decomposable into separate elements only the physical and chemistry outside of it, that which its equivalent has formed and compensates for

84

the physical reconstruction and the results is in a new reviv-ification, at the same time transported to another world. Inside the substance of the depth, is self-reproduction, a code within the soul's DNA, figuratively speaking, a blueprint speaking of the being?

"When the chemist who had taken this undertaking, who had obtained this procedure to analyze the ashes, inferred: the blood system and its blood of this race, its content gives rise to new found laws in chemistry, law's man has yet to discover.

"Accordingly how it all started was by an injection of a mutation, a virus, which produced from generation to generation on the planet of Moiromma a new breed, a new race! —out of need and desperation, to understand the soul's metabolism, and to utilize it—in creating long-life. So, death for them was—by and large, less of a mystery, and has become more like a waking up from a nightmare, and death postponed.

"Their lifespan was 120-to 250-years before this need to lengthen it transpired; now it is 500 to 800 years. It is of course just stalling death, but since their population is like 20,000-inhabitants for a planet the size of Earth's moon, there is a necessity to prolong their lifespan, and to leave their seeds on other planets so they do not become an endangered species, or for that matter, simply eradicated from time and space, and only an imprint on some cave wall like our ancestors; you must read the journal, this being left behind, to which I have his name translated to: 'Nora Iron' who had lived in the jungles of Guatemala for some fifty-years, undercover, cannibalistic; it is all in his diary, and who's to say where he is now, perhaps back on Moiromma."

Note: This story was originally written for the book: "The Cadaverous Planets" sequel to "Out of Quantum Space and Time" but not put into it, consequently it now is part of "Worlds Beyond" which is the elder book to "Out of Quantum Time and Space". #1205/ 11-3 thru 18-2016.

DR. McGEE'S INFINITY

(FROM THE: LOS ANDES OBSERVATORY)

Dr. McGee discovered something that Earth was going to witness, that being a far-off supernovae had taken place, the explosion came at 3:00 A.M., November 18, One Million B.C., and at 3:00 A.M., November 17, 2017, it would repeat itself in a lesser form. It would, he claimed, cause a gravitational pull, and at that very moment, he was to be standing 19,000-feet above sea-level, on top of one of the highest mountains of the Andes, at the Los Andes Observatory.

And this is where he is at the present.

He knew it was coming, and he wanted to test his theory of *'Infinity.'* To be dematerialized slightly, to become less apparent in physical substance for the universe to create a double of him, that would remain within the Universe for infinity, and at the same time to be able to regenerate. What form or kind of apparatus he had, was in a carry box, he kept to himself, as to draw the magnetic pull that was needed, at the right time when the wave passed.

This he had explained to the Louisiana Space Center years in advance (excluding the device), to his colleagues, but in a different way, saying *"The gravitational pull, will allow the imprint of myself to enter, or hook on to, the Universe (its dark energy), as a result, it will constitute a written letter of my existence, —to be left in space and time. It will navigate into matter and antimatter, into nothingness, forevermore."*

In a like manner, he told his coworkers, *"There is logic to everything, this is a gift I give to the cosmos of our human existence, which may not last long if we have WWIII. And at best 1000-years if we do not colonize such places as moons and conceivably Mars…and other planets."* His second point being, *"God puts us into the universe, and then takes us out, and it is up to us—in-between—to leave something behind to tell others we once were; figuratively speaking, like our ancient ancestors, who left hand prints on cave walls, or carved rock structures high in the Andes. God does, what we can't do."*

What had taken place at 3:00 A.M., that morning on the 17th, was this: a supernova had exploded, in essence, a dead star collapsed, that once existed in time and space, it has been dead for one million years, its shock waves, or Gravitational Waves, and what is left of that impact, is now passing Earth, with a solar storm at the speed of light, today, and within the pockets of the wave are other dimensions, as expected, it has grabbed onto the inner being of McGee (via his invention), his soul which maintains the imprint of the person he is (kept neatly packed within the marrow of spinal cord), and as expected particles within this solar wind, has in essence, duplicated his existence, as it has likewise made an imprint of the higher surfaces of Earth, as it passed, but in a very shallow way for Earth has several spheres and Dr. McGee has that so called apparatus to assist him with the magnetic pull he will need to accomplish his mission; and so, from the marrow of the spine, touched by the soul, his experiment has according to the good doctor, been a success.

Now back at the Louisiana Space Center, his staff and colleagues want to know how he can prove his experiment worked, that the wave, and all its elements did as he said it would do, that his double, or shadow or whatever was out there, his second-self, his depth's imprint, would be sailing throughout the Universe. Dr. McGee, asked them to walk down Main Street with him to get a cup of coffee, and as they did, they noticed his shadow, it was polarized, divided, as if—had he been the moon—you would have gotten two halves of the moon, it left them lost in a world remote.

Unresponding silence offered proof that they had been overwhelmed, now with disbanding, they left it as an unsolved mystery, although intrigued, the scientific fraternity felt no need of definite knowledge beyond what they had witnessed, it would not be published; and Dr. McGee, agreed—regarding this, —thus, it was better to leave it in the mystery of Quantum Physics, for a later date, to which humanity would be more willing to explore.

#1159/11-17-2016

THE RED BULL OF
Catalhôyûa

((DR. HiGHTOWER'S Stone AGE SEttLEMENt) (BOS PRIMIGENIUS))

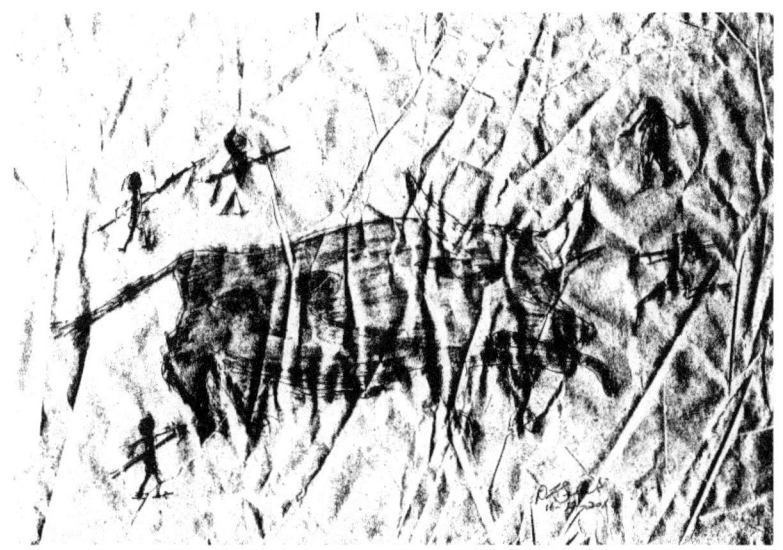

The Neolithic Red Bull of Catalhôyûa (9000 B.C.)

CHaptER OnE

"The Red Bull," **said Dr. Hightower, of the Louisiana Space Center, and directing it to the Anthropological Research Development Department, at his archeological site in Asia Minor, speaking to his colleges, updating them on his discovery** *was the size of a Blue Whale, some eighty feet long. Surely this creature to its inhabitants of that region, inspired awe; matter-of-fact, we know this for a fact, for the*

91

beast died without any wounds. It arrived in the Neolithic period of mankind (9,000 to 11,000 B.C., thereabouts). We know it devoured all the domestic plants around Catalhôyûa, in Turkey. And because of Hieroglyphics we know the beast was presumably reddish-brown, a rustic color, with black spots, long tail, an apparent outward snout, and a long hanging tongue, its feet hoofed, and its horns some six feet long. It is the only one of its kind. It would seem, the men and women of this region carried out rituals to this gigantic beast, but by feeding the bull, they per near starved themselves to death, and then the beast started feasting on them. It dominated—for the most part—the realm."

So, this was Dr. Hightower's position on the discovery, and how he inferred that the people of this region, reasoned on this ancient of hoofed quadruped.

The inhabitants proclaimed legends to the Red Bull, that it survived 1200-years in the region. And the analysts who analyzed its bones and teeth, at the Research Development Department, simply concluded, its age at 1200-years old, size at 98-feet long, and humungous weight of some, 362,000 pounds, and indicated because of having no wounds, could not tell of its demise, nor how this story ended for the men and women of this race, although all indications were that they were a bold, and powerful people. Dr. Hightower, concluded, by saying: *"Be at the conference meeting tomorrow, I will unfold the mystery."*

CHAPTER TWO

Before we go onto Dr. Hightower's unfolding of the mystery of the death of this huge beast, let me update the reader on the Catalhôyûa civilization. Their houses were built in a manner that suggest that defense was the reason for their peculiar construction. But what enemy did they have other than at times the very bull they revered. Let me explain: their dwellings had a sole entry through the roof, and their dwellings were superimposed buildings on a plain of some thirty acres, or less. One structure on top of another. Flat roofs, for this was also their walkway. Each house having a wooden stairway.

Explained, Dr. Hightower to his students, facility, and workmen and women that had been at the archeological site: *"The past, the present and the future behaves in a certain manner. And to come to my conclusion, one must study all the factors involved. Having said this, when heat arrives it changes the future, it is different from the past. If there is no friction—example, a pendulum can swing forever, this friction heats, at the same time loses energy, and slows down. Again let me say: friction produces heat. This is where the past, present, and future comes into play, only when there is heat. A fundamental phenomenon took place back then. But let me continue, heat passes from things that are hotter to things that are colder. Hot to cold. The reasons the Red Bull died, was because of a cold spell, the beast lost all its heat. If that makes sense. If you are asking 'why this is so' let me say: a quickly moving atom of the hot substance (in this case the bull) collided with a cold spell, and lost its energy, rather than vice versa. Basic physics, what is heat? Heat is atoms and/or molecules, in a cluster that moves quickly. Cold to the contrary, those atoms and/or molecules move slower. We could end this mystery right here and now, but ... (a pause) to repeat myself, hot moves to cold. Example, put a cold spoon in hot soup, it becomes hot. In a like manner on a cold day we lose body heat, and become cold. Thus the red Bull died in just this manner. Concluding, it had no wounds, and there was a long, very long cold spell and no food available for an extended period of time, and the inhabitants themselves, moved, migrated to a warmer climate. In retrospect, the bull had become anthropophagus, in that ate age human flesh, carcasses, but when the region was unpeopled for generations, it died on its own. The inhabitants revered the beast, like Americans do the bald eagle. Hence, they used spears only to keep the beast, at a distance."*

CHAPTER THREE

Dr. Hightower, could hear the endeavoring people in the background talking to one another, in an extraordinary tone of voices, asking *"How on earth did just one specie come to exist?"* **It was very painful to listen to, for honestly he knew or had an idea, and it wasn't at all keen. Said Dr. Hightower,** *"The full legend goes something like this: that the Red Bull, came from a miniature world or moon, the sphere that dropped off this beast was perfectly round in diameter, when it approached the ground a circular trapdoor opened in the bottom, several creatures descended forthwith and proceeded to their new environment, at which time they were perhaps four feet in height, three died of toxic insect bites, three vanished, or disappeared from human observation, and one survived. So the legend goes."*

#1119/11-19-2016

94

THE BABYLONIAN 12ᵀᴴ PLANET

((NIBIRU) (THE DESTROYER EXOPLANET))

Forty worldwide volcanos were all suddenly activated, starting the month of October, 2019, NASA, and the Governments worldwide kept silent on what was in the background of this phenomenon; that being, an astrophysical secret to the public. At the same time, interstellar discs, UFO spacecrafts circled the earth daily, as well as the object that had past Jupiter some three years past, and was nearing the 4th Planet, Mars.

From the volcanos, lava was pouring out into the seas, earthquakes were felt nearly every day someplace on Earth, and even the crust of the earth give the impression to tremble. It would have appeared to an observer, Earth had fallen ill.

A giant planet, ten times the size of Earth, was on its 3,600-year orbit around Earth's sun, called Nibiru (the planet of crossing), which was part of the system of Planet X, otherwise known as: *"The Dark Red Star"* or Nemesis, that carried with it within its orbit, Helion, another object within the system. The giant Exoplanet, as it was referred to at the Louisiana Space Station, by Dr. Hightower, superintendent, said: *"It is a system that moves at an angle, toward the South Pole, it will not hit Earth I believe but with its gravitational pull, who's to say what damage it will cause, and no nuclear blasts will move it one iota."*

It seemed each day a new volcano exploded, and there was great panic that prevailed, and peril, in the South Pacific Islands, throughout Bali, and Java, Indonesia, and up and down the west coast of North and South America earthquakes, beyond the 6.5 Richter scale measurement. And all this time, debris graying the skies, from: Asia, to the Americas, and across Africa, especially Japan, Mexico, along the Nepal, and Tibet boarder. And the orb continued to move, matter of fact, there was no movement of humanity within these areas, other than the natural drifting of volcanic debris, detectable.

At length after some hesitation, Russia and the United States, tried to put together a plan, calling Dr. Hightower in for discussions of what could be done, after their vain attempt of firing several nuclear missiles at the giant orb, and learning what Dr. Hightower had already explained months before, it was of no use, had it been an asteroid, perhaps.

Said Dr. Hightower, with the Chiefs of Staffs, at the White House, *"We need a milder compound of solidified gasses to sidestep the planet, and move the fabric of space in front of it, that it is going to be in, its thickness is awkward we need to wobble the planet on its axis, let it reconstruct itself by itself, after the wobble which will serve in creating a bizarre unidentifiable movement of the mass, out of its near direct orbit towards Earth. Should we not do this, we will have no crust or perhaps no longer an atmosphere, and we'll all be flying to the moon, what's to stop us?"*

It all sounded strange, and unfamiliar to the other scientists at the meeting. A Journalist from the Fox Media asked, *"Can we now tell the public of this perhaps catastrophe heading our way?"*

Said the President, *"Let's wait and see how Dr. Hightower's plan works!"*

Continued Dr. Hightower, explaining what had to be done, and the circumstances that surrounded his theory, *"Space is not continuous, although made up of grains and atoms, creating what we call space, the very thing Exoplanet is surrounded by, enmeshed in. Space in essence is a basket of atomic nuclei, looped or linked to one another from a network or fabric, or texture if you will, to/of space, likened to the way my mother used to woven her Afghans. Hence, this is what space is made up of in a nutshell, individual quanta of gravity. You slant it, make a curve, it goes with the slant (with gravity). So flows time, regardless of things, grains of space, equals matter, and space has no 'time' meaning time is absent, as well as change being ubiquitous. What we must do is change the rhythm, a single tempo in space, this will do it: the space the Exoplanet resides in or on, make it dance when it reaches that point—figuratively speaking, this will cause a quantum event. And I've already explained what the source will be, 100-rockets filled with 'solidified gasses,' to circle the planet, to disrupt the fabric of space, to create a dent as it moves into this area, so the planet will go with gravity, bend to the slant, if that makes sense."*

And so it was, from the White House to the Kremlin, from the English Parliament, to France's politicians in backrooms, to: NASA the frightening data, as it came out of the Louisiana Space Station, that this legendary 12ᵗʰ Planet, known as *'The Babylonian 12ᵗʰ Planet'* **was now crossing over Earth's solar system, and was now beyond Mars and about to pass Earth, establishing its encounter. And the world was told of its per near collision, thereafter; after Dr. Hightower's experiment worked.**

What had taken place was this, inferred Dr. Hightower to the media, after the fact: *"For a moment the Exoplanet became incapacitated, then ameliorated, — in consequence, doing what I had expected, it had shifted its angle, thus diverging from its common line, or position, from the object in front of it, Earth."*

"It's not all over yet," **said Dr. Hightower to the President, in his Oval Office,** *"It is called 'The Beginning of Sorrows' that is, in the year 2029 A.D., a decade from now, an asteroid by the given name of: Apophis, has already been spotted behind the Exoplanet, and may very well strike Earth, or if not strike Earth, move its poles, by way of its gravitational pull. There are legends among the Inuit, Eskimos, in Qikiqtaaluk, a region in Canada, of this polar move, it might be wise to get ready."*

Information take from an Article the author did on Nibiru, #1172/Oct 7, 2016/
And now made into a short SF story, 11-21 & 22-2016. #1220

Awards Given to the Author

Works by the Author

Back of Book

AWARDS TO THE AUTHOR

Siluck's Sketch Places Second In Art Show

Washington art is exhibited some place besides the hall across from 310 and senior Dennis Siluk has a plaque to prove it.

Dennis, or Chick as he is known, won second prize in the drawing division of the 9th annual "Best 100 Art Show" presented by the St. Paul Junior Chamber of Commerce and the St. Paul Art Center.

The yearly exhibit is open to young people of the St. Paul area public, private, and parochial junior and senior high schools. This year the showing dates were from May 4 to May 23 at the Art Center.

Chick's 14" x 18" chalk drawing was not the only winning entry from Mr. Rodney Magnuson's art students. Ron Anderson, Prexie junior, won honorable mention as did Tom Baier and Mary Forliti, both seniors.

Dennis Siluck

News clipping from
Washington High School
Newspaper, 1965
—Dr. Siluk's First Award—

Recognition given by the Congress of the Republic of Peru, from "DestAcados" Magazine, to Dr. Dennis L. Siluk for "Promoter of the Culture of the Mantaro Valley of Peru," May 2016.

"The Paris Book Festival" May 2016, Honorable Mention, for a Spiritual Book, completion, went to: "The Galilean" by Dr. Dennis L. Siluk.

Poet Laureate of Canchayllo, Province of Jauja, recognized by the District Municipality of Canchayllo, Peru. January 2012 (1,774-inhabitants).

Poet Laureate of Nueve de Julio, Province of Concepcion, recognized by the District Municipality of Nueve de Julio, Peru. September 2011 (3,500- inhabitants).

Poet Laureate of Satipo (Central Jungle of Peru), recognized by the Provincial Municipality of Satipo. July 2011 (21,000-inhabitants).

Poet Laureate of Huancayo (Capital of Junín), recognized by the Provincial Municipality of Huancayo, Peru. June 2011 (430,000- inhabitants).

Poet Laureate of Chilca, recognized by the District Municipality of Chilca, Junin, Peru. May 2011 (75,000-inhabitants).

Poet Laureate of Cerro of Pasco, recognized by the Provincial Municipality of Pasco, Peru. November 2007 (125,000-inhabitants).

Poet Laureate of the Mantaro Valley of Peru, recognized by the Professional Association of Journalist of Junín-Huancavelica. August 2007 (1,000,000-inhabitants).

Poet Laureate of San Jerónimo of Tunán, recognized by the District Municipality of San Jerónimo de Tunán, Peru. January 2006 ((and of the Gran Gold Cross) (10,000- inhabitants)).

Diploma of Honor by the Broadcaster Association of Central Peru, being recognized as Master of the Broadcasting on December 8, 2007 *(for several months the author had a Radio Program in Radio Universitaria, "The Moment of the Poetry," in Huancayo, Peru, hosted by Lic. Eduardo Cárdenas).*

Municipality of the Province of Jauja, First Capital of Peru, with Resolution of the Mayor Office No: 535-2011-A/MPJ 19 September 2011: recognized as Illustrious Visitor for his Poem "The Ghost of the Laguna de Paca," Sabino M. Mayor Morales, Mayor of Jauja.

Dr. Dennis L. Siluk, was bestowed the Title of Doctor Honoris Causa by the National University of Central Peru, (U.N.C.P.) on January 30, 2012, for his cultural, social, and humanitarian works on Peru.

Recognized as "Outstanding Writer of the Year" with Diploma of May 2, 2013, by the Peruvian Chamber of Enterprising and the Press Corporación of Prensa Especializada SAC ((Gerente: José Arrieta S.) (also recognized in 2007, 2011, and 2012)) with the Congress of the Republic of Peru, Lima, Peru.

Recognized as Doctor Honoris Causa by the Iberoamerican Counsel in Honor of the Leaders of Leaders (CIHLL) April 13, 2013, with its President Dr. Gladys M. Miguel Villar (Countries of South America, Central America, Spain and Portugal).

Appreciation by Pope Francis, concerning "The Galilean" through Letter, sent by the Apostolic Nunciature, Lima Peru, 2013.

Diploma of Honor by the Parish-Church of San Daniel Comboni of Cristo Redentor (Parroquia: Diócesis de Lurín) March 29, 2014.

"Dr. Dennis L. Siluk...The Order of the Legion of Mariscal Caceres... because of your excellent work in 2013, and high spirit with the people of this region, Junin (Peru), decorates you... (Ceremony to be held, 4 February 2014)" —Alejandrina Cervantes Zúñiga, President OLMC-FZRJ (Ltr. 22/01/2014).

103

Appointed, by the U.S. Embassy, Peru, Dr. Dennis L. Siluk, Warden for the City of Huancayo, Peru, 4-7-2014.

Letter, dated, 31 October 2014, to Doctor Dennis Lee Siluk: Metropolitan Municipality of Lima - Officio N° 549-14 MNL-GCSRP-SEP: '...special acknowledgement from the city for your valuable and important intellectual contribution that allows the central region of Peru, especially the Mantaro Valley...' Maria Paz Ortiz de Zevallos Subgerente de Eventos y Protocolo.

Recognition given by the Congress of the Republic of Peru, from "DestAcados" Magazine, to Dr. Dennis L. Siluk, for "Promoter of the Culture of the Mantaro Valley of Peru," 17 April 2015.

"To Mr. Dennis Siluk, in appreciation of the editing of his book 'The Magic of the Avelinos,' which is a very valuable poetic work and cultural contribution to the Mantaro Valley, Huancayo—Peru." December 1st. 2006 (The Language Center of the Peruvian University Los Andes, gives a: 'Recognition Award'). Magister Juana Andamayo Flores Coordinator, and Vice Rector Administration, Ing. Vidal Fernandez Sulca.

Decorated Vietnam War Veteran (1971) Three times decorated by the U.S. Army: Vietnam (1971); West Germany (1976); Fort Rucker, Alabama U.S.A., (1979).

United States Marine Corps Reserve: Awarded Dr. Dennis L. Siluk, Certificate of Appreciation: "Toys for Tots" for: "Outstanding Support" for giving 500-free books to the city's children; January 25, 1988 (St. Paul, Minnesota).

1965-Art Award (St. Paul, Minnesota, U.S.A.), 2nd Place: "Best 100 Art Show 1965" (J.C.s).

Etc. etc. etc...

WORKS BY THE AUTHOR

Books out of Print

The Other Door ((Poems- Volume I, 1981) (750-copies, 450 to 500-signed))

Willie the Humpback Whale ((poetic tale, 1982-83) (1st printing, 100-copies—1982; second printing, 100-copies—1983; third printing 5000-copies--1983; in 2008, 1st Spanish Version, 1000-copies printed))

The Tale of Freddy the Foolish Frog (((1982) (100 copies printed))

The Tale of Teddy and his Magical Plant (((1983) (100 copies printed))

The Tale of the Little Rose's Smile (((1983) (100-copies printed))

The Tale of Alex's Mysterious Pot (((1984) (100 copies printed))

Two Modern Short Stories of Immigrant Life [1984] 100-copies printed

The Safe Child/the Unsafe Child (((1985) (for teachers of Minnesota Schools)) 200-copies printed.

Some of the 26-Chapbooks for Peru (of the 38,000-printed)

Footprints to the Mantaro Valley ((January 2006 & February) (two printing, 50-copies of each printing signed))

A Four Part Poem for the Inauguration of the Statue Virgin Mary, of the City of Conception (100-copies signed & printed) 10-2006

The Road to Unishcoto ((9/2006) (500-copies signed))

The Poetry of Stone Forest ((9/2007) (500-copies signed))

The Magic of the Avelinos ((8/2006) (100-signed copies)) First Printing

The Legend of the Ghoul of the Laguna de Paca ((2nd Printing 2011) (1000-copies)) ...the First Printing was 50-signed copies, 2006.

El Monstrous Arcaico ((front cover title) (1000-copies, 9/2008))

The Legend of Huallallo ((2011) (booklet)) 1000-copies printed

Poetry of the Miners (2011/booklet) 1000-copies printed

Satipo, Eyebrow of the Jungle (Poems...out of Peru) 1000-copies printed

The Hidden Haven (Poems out of the Andes) 1000-copies printed

Selected Poems out of the Mantaro Valley (1/2012-1000-copies) chapbook

Peruvian Earth (Poems out of Peru) Chapbook form, (11-Poems) due 2013 (1000-copies) 23rd Peruvian Chapbook

Night over Peru ((1000-copies) (13-poems)) 26-signed, numbered and dated by the author 27th Peruvian book

The Light Around Peru (30th Chapbook, 1000-copies to be printed, a work in progress, 15-poems) 2/2016.

In The Shadows of Peru (31st Chapbook, under construction; 11-poems)?

The Universe in Motion (One Narrative Poem, 33nd Chapbook) September 2016

The Galilean I (24th Chapbook) 2013-February (1000-copies)

The Galilean II (25th Chapbook) 2013-July/1000-copies (25 signed and dated)

The Galilean III: Women Touched by God ((26th Chapbook/2013-October/1000-copies) (25 signed and dated))

The Galilean IV: The Gathering (27th Chapbook, 1000 copies) 25-signed and dated

The Galilean V: Focus on Christ (28th Chapbook, 1000-copies) 25 signed and dated

The Creation Account (29th Chapbook, 1000-copies) 25-signed and dated

The Light Around Peru (30th Chapbook, 1000 copies were printed)

The Poetic Macabre Chapbook Collection
Each chapbook's printing was between 50 to 100 copies printed

Dark Dancing Spiders (Jan. 2005) 50-signed copies printed

Legend of the Great Jaguar Beasts of Teotihuacán (Feb. 2005) 50-signed copies printed

The Lighthouse near Reykjavik (Feb. 2005) 50-signed copies printed

Things that are Dark (Jan. 2005) 50-signed copies printed

Strange Nights (Jan. 2005) 100-signed copies printed

The Age of Light (April 2005) 50-signed copies printed

The Lotus Demon of Mercury (Feb. 2005) 100-signed copies printed

The Last King of Mars (2003, never made into a chapbook) Draft

Copan (2004, never made into a chapbook) Draft

Presently in Print

Visions, Theological, Religious and Supernatural

The Last Trumpet and the Woodbridge Demon (2002) 400-copies printed

Angelic Renegades & Rephaim Giants (2002)

Islam, in Search of Satan's Rib (2002)

The Galilean (Volumes I thru VII) (2015)

The Nazarene (March 2018)

Tales of the Tiamat [trilogy]

Tiamat, Mother of Demon I (2002)

Gwyllion, Daughter of the Tiamat II (2002)

Revenge of the Tiamat III (2002)

The Addiction Books of D.L. Siluk:

A Path to Sobriety I (2002)

A Path to Relapse Prevention II (2003)

Aftercare: Chemical Dependency Recovery III (2004)

Alcoholism: Curse of the Devil ((1 Edition) (Spanish)) 9-2016

Alcoholism: Curse of the Devil ((2nd Edition, Illustrated) (Spanish)) 11-2016

Autobiographical

A Romance in Augsburg I (2003)

Romancing San Francisco II (2003)

Where the Birds don't Sing III (2003)

Stay Down, Old Abram IV (2004)

Chasing the Sun [Travels of D.L Siluk] (2002)

The Nonfiction: Short Stories Series

In My Time (((15-short stories) (Paperback & eBook)) 8-2016

Men Among Men ((Sequel to: 'In My Time') (16-short stories) (Non-fiction)) 9-2016

Time and Seasons (The Last Sixteen Tales) 9-2016

A Way You Have to Be (18-short stories) 9-2016

An Unsuspecting Life! (Book Five/Seventeen Short Stories) 11-2016

To Want, and Want Not! (Book Six/Seventeen Short Stories) 10-2016

Winner or Loser Takes Nothing (Book Seven, in the Short Story NF Series) 10-2016

A Changeable Banquet (Book Eight, in the Short Story NF Series) 10-2016

Nowhere into Nothing (Book Nine, in Siluk's Short Story NF Series) 11-2016

Things Come to Their End (Book Ten, in Siluk's Short Story NF Series) 02-2017

New Short Novels

The Protagonist (A Novelette) October 2016

Old Josh (A Short Novel) October 2016

Susanna of Bethsaida (02-2017)

Susanna of Bethsaida Bilingual ((English and Spanish) (05-2017))

Susanna of Bethsaida Bilingual 2nd. Edition ((English and Spanish) (06-2017))

Romance and/or Tragedy:

The Rape of Angelina of Glastonbury 1199 AD (2002) Novelette

Perhaps it's Love (Minnesota to Seattle) 2004 Novel

Cold Kindness (Dieburg, Germany) 2005 Novelette

Suspense, Short Stories, Novels and Novelettes:

Death on Demand (Seven Suspenseful Short Stories) 2003 Vol: I

Dracula's Ghost (And Other Peculiar Stories) 2003 Vol: II

The Jumping Serpents of Bosnia (Suspenseful Short Stories) 2008 Vol: III

The Mumbler [Psychological] 2003 (Novel)

After Eve [A Prehistoric Adventure] (2004) Novel

Mantic ore: Day of the Beast ((2002) (Novelette)) Supernatural

Every Day's Adventure ((2002) (Short Stories, etc.))

Science Fiction Short Stories

Out of Quantum Space and Time (19-short SF Stories) 9-2016

Worlds Beyond (23-short SF Stories) 11-2016

The Dark Side of Polaris ((Sequel, 12-2016) (Episodic Two Story Novelette))

The Cadaverous Planets ((An Episodic Novel of 14-SF Stories) (originally written between: 2004-2008)) 2017?

The Hyperborean Mythos ((Strange Tales) (2017?))

The Poetry of D.L. Siluk

General and Specific Poetry

The Other Door (Poems- Volume I, 1981)

Sirens (Poems-Volume II, 2003)

The Macabre Poems (Poems-Volume III, 2004)

The Tale of Willie the Humpback Whale (and two short stories) 8-2016

Specific Poetry

The Last Trumpet (2002; Prophecy in Poems) 400-printed; 100-signed.

Stone Heap of the Wildcat ((2010) (Israeli Poetry))

Last Autumn and Winter (Minnesota Poems, 2006)

Days (Poetry on Grieving: 2014)

The Galilean (Christian Poetry—87-poems) 2015

Feast of the Wolfhound (Alexandrian Epic) Paperback & eBook, 8-2016

The Eldritch Collection (Seventeen Poems) 9-2016

Ebony in Eden (Eldritch Poetry, 26-poems) Expected publication, 9-2016

Stars over Germany ((Several years in the making) (26-poems, 03/2017))

The Peruvian Collection of Poetry

Spell of the Andes [2005] (Poems out of Andes of Peru)

Peruvian Poems [2005] (Poems out of Peru)

Poetic Images out of Peru (And Other Poems, 2006)

The Magic of the Avelinos (Poems on the Mantaro Valley, Book One; 2006)

The Road to Unishcoto (Poems on the Mantaro Valley, Book Two, 2007)

The Poetry of Stone Forest (Cerro de Pasco, 2007)

The Windmills (Poetry of: Juan Parra del Riego) 2009

The People Will Not Break ((Peru) (Poems out of the Mantaro Valley, Peru)) 2012

The Natural Writings of D.L. Siluk

Cornfield Laughter (and the unpublished collected stories...) 2009 (Vol. I) 300 pp

Men with Torrent Women (Two Short Novelettes and Sixteen Short stories) 2009 (Vol. II) 250 pp

A Leaf and a Rose (a comprehensive library of new writings...) 2009, (Vol. III) 500 pp

The Cotton Belt ((An Episodic Novel of the Old South) (Volume IV)) 2011/616 pp

Plays

Six Plays (June 2017)

Artwork

The Artwork of Dr. Dennis L. Siluk, Volume I (August 2017)

The Artwork of Dr. Dennis L. Siluk, Volume II (February 2018)

Works in Progress

The Times I Live In! (Book Seven/Eleven Commentaries on the times)

Back of Book

The author working on two of his books, November, 2016, on his roof top patio, in Lima, Peru.

In these twenty-three short stories of Science Fiction, called *'Worlds Beyond'* comes quantum visions of infinite spheres and multiple dimensions. The book similar to *"Out of Quantum Space and Time,"* but more of a reworked version, with four additional stories, thus, not a 2nd Edition, but first in its own right. Brilliantly conceived and executed in a style that is unique for the 21st Century reader, with over twenty illustrations. There is no imitativeness, all are original in concept and performance. The readings are a discovery, a quick read, entertaining, and astonishingly.

Dr. Siluk, brings together: quantum physics, demonology, archeology, anthropology, genetics, astrology, biology, geology, eschatology, lunar geology, geochemistry, necromancy, neo-anthropophagus, anthropomorphism, hieroglyphics (rock art), epistemology, and psychology.

This is Dr. Siluk's 65th international book, "Worlds Beyond." He is a poet since he was twelve years old, a writer, Psychologist, Ordained Minister, Decorated Veteran from the Vietnam War, Doctor in Arts and Education. In addition, he received twice Honorary Doctorate, and was appointed Poet Laureate in Peru nine times. One of his books, "The Galilean," took Honorable Mention at the 2016 Paris Book Festival, and received an award from the Congress of Peru, for his cultural writings. He is originally from St. Paul, Minnesota, and lives with his wife Rosa, in Lima, Peru and High up in the Andes, in Huancayo, also, in Minnesota.

Back picture on the book cover is of the Author on Easter Island, 2002.

www.ingramcontent.com/pod-product-compliance
Lightning Source LLC
Chambersburg PA
CBHW070030210526
45170CB00012B/528